Ben Stacy Jerrik (Ed.)

MS Stena Superfast VIII

Ben Stacy Jerrik (Ed.)

MS Stena Superfast VIII

Tallinn, Helsinki, Rostock, Paldiski, MS Star, Stena Line

Part Press

Imprint

Publisher:
Part Press is a trademark of
International Book Market Service Ltd., 17 Rue Meldrum, Beau Bassin, 1713-01 Mauritius
Email: info@bookmarketservice.com
Website: www.bookmarketservice.com

Published in 2012

Printed in: U.S.A., U.K., Germany. This book was not produced in Mauritius.

ISBN: 978-620-1-66130-1

Contents

Articles

References

'

MS *Superfast VIII* near Helsinki, June 2007.

Career	
Name:	***Stena Superfast VIII***
Owner:	2002—2006: Superfast Ferries 2006 onwards: Tallink[1]
Operator:	2002—2006: Superfast Ferries 2006-2011: Tallink 2011 onwards: Stena Line[1]
Port of registry:	2002—2006: Piraeus, Greece 2006-2011: Paldiski, Estonia 2011 onwards: Belfast, United Kingdom[1]
Route:	Belfast—Cairnryan
Builder:	HDW, Kiel, Germany
Yard number:	358[1]
Launched:	28 November 2000[1]
Acquired:	11 July 2001[1]
Maiden voyage:	15 July 2001[1]
In service:	15 July 2001[1]
Identification:	IMO number: 9198953[1]
Status:	In service
General characteristics (as built)[1]	
Class and type:	*Superfast VII* class fast ropax ferry
Tonnage:	30,285 GT
Displacement:	5990 t (**unknown operator: u'strong'** long tons) deadweight (DWT)
Length:	203.3 m (**unknown operator: u'strong'** ft)
Beam:	25.42 m ()
Draught:	6.60 m ()

Decks:	10[2]
Ice class:	1 A Super[2]
Installed power:	4 × Wärtsilä-Sulzer NSD ZA V40S diesels combined 46000 kW (**unknown operator: u'strong'** shp)
Speed:	30.4 kn (**unknown operator: u'strong'** km/h)[2]
Capacity:	626 passengers 661 cars 1920 lane meters

MS *Superfast VIII* is a fast Ro-Pax ferry owned by the Estonian ferry company Tallink servicing the route Tallinn (Estonia) – Helsinki (Finland) – Rostock (Germany). Built in 2001 by Howaldtswerke-Deutsche Werft (HDW), Kiel, Germany for Attica Group's subsidiary Superfast Ferries, she was sold to her current owners in 2006.[1]

In March 2006 Superfast sold their Baltic Sea operations to Tallink. Like her sisters, the Superfast VIII was moved from Finnish to Estonian registry, and her route changed to Hanko—Paldiski—Rostock from April 2006 onwards. The call at Paldiski proved to be impractical, an already in June of the same year the route reverted back to Hanko — Rostock. Around this time the ship's hull marking where changed to "Superfast operated by Tallink". Although there were rumours that the ships would be moved under Silja Line's brand from January 2007 onwards, they were instead officially made a part of Tallink's fleet. At the same time their route changed to Tallinn—Helsinki—Rostock. Between 5 and 12 April 2007 the ship was used on the Helsinki—Tallinn route due to a delay in the delivery of the new MS *Star*. After this she joined her sisters in the Rostock service.

In March 2011, Stena Line announced they will be chartering the *Superfast VIII* and sister ship *Superfast VII*. The vessels will operate between Belfast and Cairnryan.[3]

References

[1] Asklander, Micke. "M/S *Superfast VIII* (2001)" (http://www.faktaomfartyg.nu/superfast_VIII_2001.htm) (in in Swedish). *Fakta om Fartyg*. . Retrieved 2008-04-02.

[2] "Tallink Superfast » Technical information" (http://www.tallinksilja.com/en/tab2/ships/superfast/technicalInformation/). Tallink official website. . Retrieved 2008-04-02.

[3] "New ships for Scotland - NI service" (http://www.stenaline.co.uk/ferry/media/news/newships/). Stena Line. 7 March 2011. .

External links

- M/S Superfast VIII at Fakta om Fartyg (http://www.faktaomfartyg.nu/superfast_VIII_2001.htm) (in Swedish)
- Superfast VIII at marinetraffic.com (http://www.marinetraffic.com/ais/shipdetails.aspx?mmsi=235089436)

Tallinn

Tallinn

From top-left: Guard towers along the city wall, Kadriorg Palace, St Mary's Cathedral and Town Hall spire, KUMU, Tallinn's modern skyline, Panorama of the Old Town.

Flag

Coat of arms

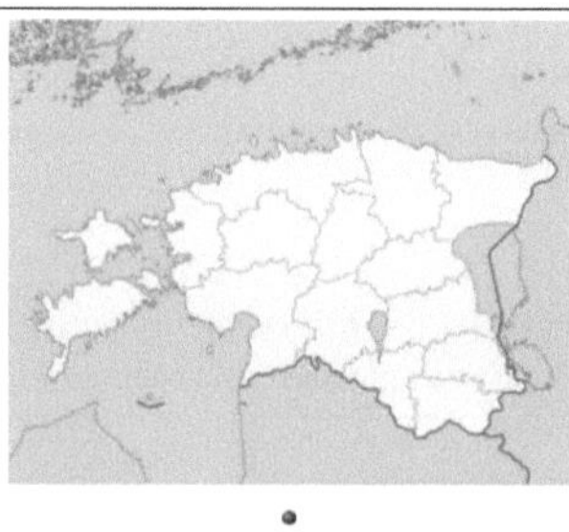

Tallinn

Coordinates: 59°26′14″N 24°44′43″E

Country	Estonia
County	Harju County
First appeared on map	1154
Town rights	1248
Government	
• **Mayor**	Edgar Savisaar (Centre Party)
Area	

• **Total**	159.2 km^2 (61.5 sq mi)
Population (1 June 2012[1])	
• **Total**	416,536
Time zone	EET (UTC+2)
• **Summer (DST)**	EEST (UTC+3)
Website	www.tallinn.ee [2]

Tallinn (/ˈtɑːlɪn/ or /ˈtælɪn/,[3] [4] Estonian pronunciation: [ˈtɑlʲˑinˑ]) is the capital and largest city of Estonia. It occupies an area of 159.2 km^2 (**unknown operator: u'strong'** sq mi) with a population of 416,536.[1] It is situated on the northern coast of the country, on the shore of the Gulf of Finland, 80 km (**unknown operator: u'strong'** mi) south of Helsinki, east of Stockholm and west of Saint Petersburg. Tallinn's Old Town is in the list of UNESCO World Heritage Sites. It is ranked as a global city and has been listed among the top 10 digital cities in the world.[5] The city was a European Capital of Culture for 2011, along with Turku, Finland.

Tallinn is the oldest capital city in Northern Europe.[6] [7] The city was known as **Reval** from the 13th century until the 1920s.

Toponymy

Historical names

The lesser coat of arms of Tallinn is also the coat of arms of Harju County and depicts the Dannebrog cross.

In 1154 a town called *Qlwn*[8] or *Qalaven* (possible derivations of *Kalevan* or *Kolyvan*)[9] [10] was put on the world map of the Almoravid by the Muslim cartographer Muhammad al-Idrisi, who described it as *a small town like a large castle* among the towns of *Astlanda*. It has been suggested that the *Quwri in Astlanda* may have denoted the predecessor town of today's Tallinn.[11] [12]

The earliest names of Tallinn include *Kolyvan* (Russian: Колывань) known from East Slavic chronicles, the name possibly deriving from the Estonian mythical hero *Kalev*.[13] [14]

Up to the 13th century the Scandinavians and Henry of Livonia in his chronicle called the town Lindanisa: *Lyndanisse* in Danish,[15] [16] [17] *Lindanäs* in Swedish, also mentioned as *Ledenets* in Old East Slavic. According to some theories the name derived from mythical Linda, the wife of Kalev and the mother of Kalevipoeg.[18] who in an Estonian legend carried rocks to her husband's grave that formed the Toompea hill.[19]

It has been also suggested that in the context the meaning of *linda* in the archaic Estonian language, that is similar to *lidna* in Votic, had the same meaning as *linna* or *linn* later on meaning a castle or town in English. According to the suggestion *nisa* would have had the same meaning as *niemi* (meaning peninsula in English) in an old Finnish form of the name *Kesoniemi*.[20]

Other than *Kesoniemi* known ancient historical names of Tallinn in Finnish include *Rääveli*. The Icelandic Njálssaga mentions Tallinn and calls it Rafala, which is a variant of the name Raphael.

After the Danish conquest in 1219 the town became known in the German, Swedish and Danish languages as Reval (Latin: *Revalia*). The name originated from (Latin) *Revelia* (Estonian) *Revala* or *Rävala*, the adjacent ancient name of the surrounding Estonian county.

Modern name

The origin of the name "Tallinn(a)" is certain to be Estonian, although the original meaning of the name is debated. It is usually thought to be derived from "*Taani-linn(a)*" (meaning "Danish-castle/town"; Latin: Castrum Danorum) after the Danes built the castle in place of the Estonian stronghold at Lindanisse. However, it could also have come from "*tali-linna*" ("winter-castle/town"), or "*talu-linna*" ("house/farmstead-castle/town"). The element *-linna*, like Germanic *-burg* and Slavic *-grad* / *-gorod*, originally meant "fortress" but is used as a suffix in the formation of town names.

The previously used official German name **Reval** (*Ревель*) was replaced after Estonia became independent in 1918–1920. At first both forms *Tallinna* and *Tallinn* were used.[21] The United States Board on Geographic Names adopted the form *Tallinn* between June 1923 and June 1927.[22] The form *Tallinna* appearing in modern times in Estonian denotes the genitive case of the name, as in *Tallinna Reisisadam* (*Port of Tallinn*).

Other variations of modern spellings include **Tallinna** in Finnish and **Та́ллин** in Russian.

A form **Tallin** deriving from the Romanization of Russian spelling of the name *Та́ллин*[23] was also used internationally during the era Estonia was annexed to the Soviet Union.

History

Historic Centre (Old Town) of Tallinn *	
UNESCO World Heritage Site	

Country	Estonia
Type	Cultural
Criteria	ii, iv
Reference	822bis [24]
Region **	Europe and North America
Inscription history	
Inscription	1997 (21st Session)
* Name as inscribed on World Heritage List [25] ** Region as classified by UNESCO [26]	

The Danish flag falling from the sky in the 1219 Battle of Lyndanisse.

The old town

Seal of Tallinn, 1340

Port of Tallinn in 1853

The first traces of human settlement found in Tallinn's city center by archeologists are about 5000 years old. The comb ceramic pottery found on the site dates to about 3000 BC and corded ware pottery c. 2500 BC.[27]

In 1050 the first fortress was built on Tallinn Toompea.[9]

As an important port for trade between Russia and Scandinavia, it became a target for the expansion of the Teutonic Knights and the Kingdom of Denmark during the period of Northern Crusades in the beginning of the 13th century when Christianity was forcibly imposed on the local population. Danish rule of Tallinn and Northern Estonia started in 1219.

In 1285 the city became the northernmost member of the Hanseatic League – a mercantile and military alliance of German-dominated cities in Northern Europe. The Danes sold Tallinn along with their other land possessions in northern Estonia to the Teutonic Knights in 1346. Medieval Tallinn enjoyed a strategic position at the crossroads of trade between Western and Northern Europe and Russia. The city, with a population of 8,000, was very well fortified with city walls and 66 defence towers.

A weather vane, the figure of an old warrior called Old Thomas, was put on top of the spire of the Tallinn Town Hall in 1530 that became the symbol for the city.

With the start of the Protestant Reformation the German influence became even stronger as the city was converted to Lutheranism. In 1561 Tallinn politically became a dominion of Sweden.

During the Great Northern War, Tallinn along with Swedish Estonia and Livonia capitulated to Imperial Russia in 1710, but the local self-government institutions (Magistracy of Reval and Chivalry of Estonia) retained their cultural and economical autonomy within Imperial Russia as the Duchy of Estonia. The Magistracy of Reval was abolished in 1889. The 19th century brought industrialization of the city and the port kept its importance. During the last decades of the century Russification measures became stronger.

On 24 February 1918, the Independence Manifesto was proclaimed in Tallinn, followed by Imperial German occupation and a war of independence with Russia. On 2 February 1920, the Tartu Peace Treaty was signed with Soviet Russia, wherein Russia acknowledged the independence of the Estonian Republic. Tallinn became the capital of an independent Estonia. After World War II started, Estonia was occupied by the Soviet Union (USSR) in 1940, and later occupied by Nazi Germany from 1941 to 44. In March 44 Tallinn got heavily damaged in Soviet Union air raid

(March bombing). After the Nazi retreat in 1944, it was again occupied by the USSR. After annexation into the Soviet Union, Tallinn became the capital of the Estonian SSR.

During the 1980 Summer Olympics, the sailing (then known as yachting) events were held at Pirita, north-east of central Tallinn. Many buildings, such as the "Olümpia" hotel, the new Main Post Office building, and the Regatta Center, were built for the Olympics.

The original Old Thomas (1530)

In August 1991 an independent democratic Estonian state was re-established and a period of quick development to a modern European capital ensued. Tallinn became the capital of a de facto independent country once again on 20 August 1991.

Tallinn has historically consisted of three parts:

- The Toompea (*Domberg*) or "Cathedral Hill", which was the seat of the central authority: first the Danish captains, then the komturs of the Teutonic Order, and Swedish and Russian governors. It was until 1877 a separate town (*Dom zu Reval*), the residence of the aristocracy; it is today the seat of the Estonian parliament, government and some embassies and residencies.
- The Old Town, which is the old Hanseatic town, the "city of the citizens", was not administratively united with Cathedral Hill until the late 19th century. It was the centre of the medieval trade on which it grew prosperous.
- The Estonian town forms a crescent to the south of the Old Town, where the Estonians came to settle. It was not until the mid-19th century that ethnic Estonians replaced the local Baltic Germans as the majority amongst the residents of Tallinn.

Historically, the city has been attacked, sacked, razed and pillaged on numerous occasions. Although extensively bombed by Soviet air forces during the latter stages of World War II, much of the medieval Old Town still retains its charm. The Tallinn Old Town (including *Toompea*) became a UNESCO World Cultural Heritage site in 1997.

At the end of the 15th century a new 159 m (**unknown operator: u'strong'** ft) high Gothic spire was built for St. Olaf's Church. Between 1549 and 1625 it was the tallest church in the world. After several fires and following rebuilding, its overall height is now 123 m (**unknown operator: u'strong'** ft).

Panorama of the central Town Hall Square (Raekoja plats)

Geography

Panorama of Tallinn's City Centre

Tallinn is situated on the southern coast of the Gulf of Finland, in north-western Estonia.

The largest lake in Tallinn is Lake Ülemiste (covering 9.6 km²). It is the main source of the city's drinking water. Lake Harku is the second largest lake within the borders of Tallinn and its area is 1.6 km². Tallinn does not lie on a major river. The only significant river in Tallinn is Pirita River in Pirita, a city district counted as a suburb. Historically, the small Härjapea River flowed from Lake Ülemiste through the town into the sea, but the river was diverted for sewage in the 1930s and has since completely disappeared from the cityscape.

A limestone cliff runs through the city. It can be seen at Toompea, Lasnamäe and Astangu. However, Toompea is not a part of the cliff, but a separate hill.

The highest point in Tallinn, at 64 meters above sea level, is situated in Hiiu, Nõmme District, in the south-west of the city.

The length of the coastline is 46 kilometres. It comprises three bigger peninsulas: Kopli peninsula, Paljassaare peninsula and Kakumäe peninsula.

Climate

Administrative districts

Districts of Tallinn

District	Area	Population
1. Haabersti	18.6 km²	42,294
2. Kesklinn (centre)	28.0 km²	51,308
3. Kristiine	9.4 km²	30,055
4. Lasnamäe	30.0 km²	116,273
5. Mustamäe	8.0 km²	64,237
6. Nõmme	28.0 km²	38,898
7. Pirita	18.7 km²	16,636
8. Põhja-Tallinn	17.3 km²	56,433

For local government purposes, Tallinn is subdivided into 8 administrative districts (Estonian: *linnaosad*, singular *linnaosa*). The district governments are city institutions that fulfill, in the territory of their district, the functions assigned to them by Tallinn legislation and statutes.

Each district government is managed by an Elder (Estonian: *linnaosavanem*). He or she is appointed by the City Government on the nomination of the Mayor and after having heard the opinion of the Administrative Councils. The function of the Administrative Councils is to recommend, to the City Government and Commissions of the City Council, how the districts should be administered.

Population

Ethnic composition (2009)[30]

Ethnic group	Percentage
Estonians	52.2%
Russians	38.6%
Ukrainians	3.8%
Belarusians	2.1%
Finns	0.6%
Others	2.7%

The registered population of Tallinn is 416,536 as of 1 June 2012.[1]

According to Eurostat, in 2004 Tallinn had the largest number of non-EU nationals of all EU member states' capital cities.[31] As of 2009 around 22% of its population are not EU citizens.[30]

The dominant language of Tallinn is Estonian, followed closely by Russian. As of 2011, 206,490 (50.1%) speak Estonian as their native language and 192,199 (46.7%) speak Russian as their native language. Other spoken languages include, Ukrainian, Belarusian and Finnish.[32]

Native languages in Tallinn, 2011.[32]	
Estonian	50%
Russian	47%
Other	3%

Population development

Year	1372	1772	1816	1834	1851	1881	1897	1925	1959	1989	1996	2000	2005	2006	2007	2011	2012
Population	3,250	6,954	12,000	15,300	24,000	45,900	58,800	119,800	283,071	478,974	427,500	400,378	401,694	399,108	400,911	415,416	416,536

Economy

In addition to longtime functions as seaport and capital city, Tallinn has seen development of an information technology sector; in its 13 December 2005, edition, *The New York Times* characterized Estonia as "a sort of Silicon Valley on the Baltic Sea".[33] One of Tallinn's sister cities is the Silicon Valley town of Los Gatos, California. Skype is one of the best-known of several Estonian start-ups originating from Tallinn. Many start-ups started from the Soviet-era Institute of Cybernetics. The economic sectors of Tallinn also include the light, textile, and food industry, as well as the service and government sector. There is a small fleet of ocean going-trawlers that operate out of Tallinn.[34] Port of Tallinn is one of the biggest ports in the Baltic sea region.[35]

Tornimäe business area

Currently, over half of the Estonian GDP is created in Tallinn.[36] In 2008, the GDP per capita of Tallinn stood at 172% of the Estonian average.[37] This makes the GDP of Tallinn be at 115% of the European Union average, while the overall GDP level of Estonia is at 67% of the EU average.

Rotermanni business district

Notable headquarters

Among others:

- NATO Cooperative Cyber Defence Centre of Excellence (CCDCOE)
- European Agency for the operational management of large-scale IT systems in the area of freedom, security and justice[38] [39] will be based in Tallinn.
- Estonian Air has its headquarters in Tallinn.[40]
- Skype has its software development centre located in Tallinn.[41]
- TeliaSonera has its IT development centre located in Tallinn.[42]

- Ericsson has one of its biggest production facilities in Europe located in Tallinn, focusing on the producing of 4G communication devices.[43]
- Statoil has announced moving the group's financial centre to Tallinn.[44]

Education

Institutions of higher education and science include:

- Tallinn University
- Tallinn University of Technology
- Estonian Business School
- Estonian Academy of Music and Theatre
- Estonian Academy of Arts
- Estonian Maritime Academy
- Public Service Academy
- Estonian Evangelical Lutheran Church Institute of Theology

The sports hall of Tallinn English College

Tourism

Since its return to independence, improving air and sea transport links with Western Europe and Estonia's accession to the European Union have made Tallinn easily accessible to tourists.

Estonia has made rapid economic progress since independence and this is reflected in local prices.[45] Although not extortionate, neither are prices as cheap as in other former Eastern Bloc countries.

The Stenbock House – Residence of the Estonian government on Toompea hill

The main attractions are in the two old towns (Lower Town and Toompea) which are both easily explored on foot. Eastern districts around Pirita and Kadriorg are also worth visiting and the Estonian Open Air Museum (*Eesti Vabaõhumuuseum*) in Rocca al Mare, west of the city, preserves aspects of Estonian rural culture and architecture.

Toompea – Upper Town

This area was once a separate town (Dom zu Reval), the residence of the Chivalry of Estonia, Roman Catholic bishops of Tallinn (until 1561) and Lutheran superintendents of Estonia, occupying an easily defensible site overlooking the surrounding districts. The major attractions are the walls and various bastions of Castrum Danorum, the Russian Orthodox Alexander Nevsky Cathedral (built during the period of Russian Empire, the church was built on a site that formerly housed a statue of Martin Luther) and the Lutheran Cathedral (*Toomkirik*) and the old Estonian Royal Palace now the Parliament building.

The mixture of old and new is seen everywhere in Tallinn

All-linn – Lower Town

This area is one of the best preserved old towns in Europe and the authorities are continuing its rehabilitation. Major sights include Raekoja plats (Town Hall square), the town walls and towers (notably "Fat Margaret" and "Kiek in de Kök") and St Olaf church tower (124 m).

Kadriorg

This is 2 kilometres east of the centre and is served by buses and trams. Kadriorg Palace, the former palace of Peter the Great, built just after the Great Northern War, now houses (part of) the Art Museum of Estonia, presidential residence and the surrounding grounds include formal gardens and woodland. Restored 2001–2004 with a large donation from the Swedish Government

The new residence of the Art Museum of Estonia, KUMU (Kunstimuuseum, Art Museum), was built in 2006. It has an encyclopedic collection of Estonian Art, including paintings by Carl Von Neff, Johann Köler ("Monk's Impression of Lorelei"), Eduard Ole ("Spring"), Jaan Koort, Konrad Mägi ("Norwegian Landscape"), Eduard Wiiralt, Henn Roode and Adamson-Eric.

St. Olaf's church was the tallest building in the world from 1549 to 1625

Pirita

This coastal district is a further two kilometres north-east of Kadriorg. The marina was built for the Moscow Olympics of 1980, and boats can be hired on the Pirita River. Two kilometres inland are the Botanic Gardens and the Tallinn TV Tower.

Part of Lower Town city wall.

Kadriorg Palace

Viru Gate, entrance to the Old Town. One of two remaining towers that were once part of a larger fourteenth-century gate system.

St. Catherine's Passage

Alexander Nevsky Cathedral built in 1894–1900.

Kiek in de Kök defence tower.

St. Nicholas' Church

House of the Brotherhood of Blackheads.

Pirita Convent

Tallinn Song Festival Grounds.

Transport

City transport

The city operates a system of bus (62 lines), tram (4 lines) and trolley-bus (8 lines) routes to all districts. A flat-fare system is used. Payment for single tickets is made either by pre-purchase of tickets at street-side kiosks or by a purchase from the transport vehicle. Monthly cards are available by registering through the national ID-card.

Air

The Lennart Meri Tallinn Airport is about four kilometres from Town Hall square (*Raekoja plats*). There is a local bus connection between the airport and the edge of the city centre (bus no. 2). The nearest railway station Ülemiste is only 1.5 km from the airport.

The construction of the new section of the airport began in 2007 and was finished in summer 2008.

There has been a helicopter service to and from Helsinki operated by Copterline and taking 18 minutes to cross the Gulf of Finland. The

An Estonian Air plane in Lennart Meri Tallinn Airport.

Copterline Tallinn terminal is located adjacent to Linnahall, five minutes from the city center. After a crash near Tallinn in August 2005, service was suspended but restarted in 2008 with a new fleet.[46] The operator cancelled it again in December 2008,[47] on grounds of unprofitability. On 15 February 2010, Copterline filed for bankruptcy, citing inability to keep the company profitable. In 2011 Copterline started again operating the Tallinn-Helsinki flights.

Ferry

Several ferry operators, Viking Line, Linda Line Express, Tallink and Eckerö Line, connect Tallinn to

- Helsinki (Finland)
- Mariehamn (Finland)
- Stockholm (Sweden)
- Rostock (Germany)
- St-Petersburg (Russia)

Passenger lines connect Tallinn to Helsinki (83 kilometres north of Tallinn) in approximately 2–3.5 hours by cruiseferries.

Helicopter embarking on a route to the capital of Finland, Helsinki.

Port of Tallinn

"Baltic Princess" of ferry operator Tallink.

Ferry operator Tallink vessels and the Tallinn TV Tower in the background

Tallinn is a popular destination for cruise ships.

Rail and road

Train station in Tallinn

The Edelaraudtee railway company operates train services from Tallinn to Tartu, Valga, Türi, Viljandi, Tapa, Narva, Orava, and Pärnu. Buses are also available to all these and various other destinations in Estonia, as well as to Saint Petersburg in Russia and Riga, Latvia. The Go Rail company operates a daily international sleeper train service between Tallinn-Moscow.

Tallinn also has a commuter rail service running from Tallinn's main rail station in two main directions: east (Aegviidu) and to several western destinations (Pääsküla, Keila, Riisipere, Paldiski, Klooga and Kloogaranna). These are electrified lines and are used by the Elektriraudtee railroad company. The trains are a mixture of modernised older Soviet EMUs and newly built units. The first electrified train service in Tallinn was opened in 1924 from Tallinn to Pääsküla, a distance of 11.2 kilometres.

The Rail Baltica project, which will link Tallinn with Warsaw via Latvia and Lithuania, will connect Tallinn with the rest of the European rail network. A tunnel has been proposed between Tallinn and Helsinki, though it remains at a planning phase.

The Via Baltica motorway (part of European route E67 from Helsinki to Prague) connects Tallinn to the Lithuanian/Polish border through Latvia.

Frequent and affordable long-distance bus routes connect Tallinn with other parts of Estonia.

International relations

Twin towns – sister cities

Tallinn participates in international town twinning schemes to foster good international relations. Partners include:

- Annapolis, Maryland, United States
- Białystok, Poland
- Dartford, United Kingdom
- Gdańsk, Poland
- Ghent, Belgium
- Gothenburg, Sweden
- Groningen, Netherlands[48] [49]
- Kiel, Germany
- Kotka, Finland (1955)
- Łomża, Poland
- Los Gatos, California, United States
- Malmö, Sweden[50]
- Moscow, Russia
- Portland, Oregon, United States
- Riga, Latvia[51]
- Saint Petersburg, Russia
- Schwerin, Germany
- Stockholm, Sweden
- Venice, Italy
- Vilnius, Lithuania[52]

See also

- Eurovision Song Contest 2002
- Legends of Tallinn
- Soviet evacuation of Tallinn 1941
- Tallinn Marathon

References

[1] "Tallinna elanike arv [Number of Tallinn residents]" (http://www.tallinn.ee/g4258s9268) (in Estonian). *tallinn.ee*. Tallinn city government. 1 June 2012. . Retrieved 6 June 2012.

[2] http://www.tallinn.ee/eng

[3] "Tal•linn" (http://dictionary.infoplease.com/tallinn). Dictionary.infoplease.com. . Retrieved 2012-05-20.

[4] "Definition of Tallinn" (http://encyclopedia2.thefreedictionary.com/Tallinn). Encyclopedia2.thefreedictionary.com. . Retrieved 2012-05-20.

[5] "Tech capitals of the world" (http://www.theage.com.au/small-business/tech-capitals-of-the-world-20090619-co0t.html). Theage.com.au. 2012-05-15. . Retrieved 2012-05-20.

[6] (http://tutvustaeestit.eas.ee/en/facts-about-estonia-and-publications/positively-surprising/interesting-facts-about-estonia/), Estonian introduction materials.

[7] Marvell, Alan (2005). *GCE AS Travel and Tourism Single Award for Edexcel*. Heinemann. pp. 131. ISBN 9780435446420.

[8] Fasman, The Geographer's Library, pp.17

[9] Ertl, Alan (2008). *Toward an Understanding of Europe* (http://books.google.com/books?id=X9PGRaZt-zcC&pg=PA381). Universal-Publishers. p. 381. ISBN 978-1-59942-983-0. .

[10] Birnbaum, Stephen; Mayes Birnbaum, Alexandra (1992). *Birnbaum's Eastern Europe* (http://books.google.com/books?source=ig&hl=en&rlz=&=&q="who called the settlement Kolyvan"&um=1&ie=UTF-8&sa=N&tab=wp). Harper Perennial. ISBN 978-0-06-278019-5. .

[11] Fasman, Jon (2006). *The Geographer's Library* (http://books.google.com/books?id=bE2oerrW_IkC&pg=PA17&dq). Penguin. p. 17. ISBN 978-0-14-303662-3. .

[12] "A glance at the history and geology of Tallinn" by Jaak Nõlvak. In Wogogob 2004: Conference Materials (http://www.gi.ee/WOGOGOB/wogogob_web.pdf)

[13] Terras, Victor (1990). *Handbook of Russian Literature* (http://books.google.com/books?id=VjKh2gkCudAC&pg=PA68&dq). Yale University Press. p. 68. ISBN 978-0-300-04868-1. .

[14] *The Esthonian Review* (http://books.google.com/books?id=-D9DAAAAIAAJ&q="the+old+Russian+name+for+Reval+has+been+retained+(Kolyvan+from+Kalev)"&dq="the+old+Russian+name+for+Reval+has+been+retained+(Kolyvan+from+Kalev)"&ei=LRUISeivAaX2MaLHpJwB&client=firefox-a&pgis=1). University of California. .

[15] **(Danish)**In 1219 Valdemar II of Denmark, leading the Danish Fleet in connection with the Livonian Crusade, landed in an Estonian town of Lindanisse

[16] "Salmonsens Konversations Leksikon" (http://runeberg.org/salmonsen/2/7/0518.html). Runeberg.org. 2012-01-19. . Retrieved 2012-05-20.

[17] **(German)** Reval's ältester Estnischer Name Lindanisse, Verhandlungen der gelehrten estnischen Gesellschaft zu Dorpat. Band 3, Heft 1. Dorpat 1854, p. 46–47

[18] Wieczynski, Joseph (1976). *The Modern Encyclopedia of Russian and Soviet History*. Academic International Press. p. 230. ISBN 978-0-87569-064-3.

[19] Ransome, Arthur (1923). *"Racundra's" First Cruise* (http://books.google.com/books?q="The+old+giant+Kalev+died+here+at+Reval,+and+Linda+heaped+stone+after+stone+upon+his+grave+and+so+made+that+proud+hill"&btnG=Search+Books). B.W. Huebsc. .

[20] VIRKKUNEN, A. H. (1907) (in Finnish). *ITÄMEREN SUOMALAISET SAKSALAISEN VALLOITUKSEN AIKANA* (http://books.google.com/books?id=OK4MAAAAYAAJ&pg=PA91). Suomen Muinaismuistoyhdistys. p. 91. .

[21] Singer, Nat A.; Steve Roman (2008). *Tallinn In Your Pocket* (http://books.google.com/books?id=PZdt1EnuafsC&pg=PA13&dq). In Your Pocket. p. 11. ISBN 0-01-406269-0. .

[22] *Decisions of the United States Geographic Board* (http://books.google.com/books?id=F40tAAAAMAAJ&pg=RA8-PA39&dq). The Board.. .

[23] Young, Jekaterina (1990). *Russian at Your Fingertips* (http://books.google.com/books?id=rve6qRtMPYUC&pg=RA1-PA100&dq). Routledge. p. 100. ISBN 0-415-02930-9. .

[24] http://whc.unesco.org/en/list/822

[25] http://whc.unesco.org/en/list

[26] http://whc.unesco.org/en/list/?search=&search_by_country=&type=&media=&region=&order=region

[27] Alas, Askur. "The mystery of Tallinn's Central Square" (http://www.ekspress.ee/2008/10/29/eesti-uudised/5040-vabaduse-platsi-mysteerium-kuhu-kadus-kaks-sajandit-ajalugu) (in Estonian). EE. . Retrieved 29 October 2008.

[28] "Pogoda.ru.net" (http://pogoda.ru.net/climate/26038.htm) (in Russian). . Retrieved 7 September 2007.

[29] Climatological Information for Tallinn, Estonia (http://www.weather.gov.hk/wxinfo/climat/world/eng/europe/n_europe/tallinn_e.htm), Hong Kong Observatory accessed 1 May 2012.

[30] *Statistical Yearbook of Tallinn 2008* (http://www.tallinn.ee/est/g2677s45500). Tallinn: Tallinn City Government. 2009. p. 160. .

[31] Eurostat (2004). *Regions: Statistical yearbook 2004* (http://epp.eurostat.ec.europa.eu/cache/ITY_OFFPUB/KS-AF-04-001/EN/KS-AF-04-001-EN.PDF). Luxembourg: Office for Official Publications of the European Communities. p. 135. .

[32] "Tallinn arvudes / Statistical Yearbook of Tallinn" (http://www.tallinn.ee/est/g2677s56143) (in Estonian, English) (PDF). Tallinn City Council. 3 August 2011. . Retrieved 1 April 2012.

[33] Mark Ländler, "The Baltic Life: Hot Technology for Chilly Streets" (http://select.nytimes.com/gst/abstract.html?res=F00614F935550C708DDDAB0994DD404482), *The New York Times*, 13 December 2005.

[34] Reyktal AS fleet (http://www.iec.is/reyktal.htm)

[35] "History | Tallinna Sadam" (http://www.portoftallinn.com/history). Portoftallinn.com. . Retrieved 3 June 2011.

[36] Kaja Koovit. "Half of Estonian GDP is created in Tallinn" (http://www.balticbusinessnews.com/?PublicationId=f47e445a-e234-432f-b7b0-137ca0bff47d). Balticbusinessnews.com. . Retrieved 2012-05-20.
[37] "Half of the gross domestic product of Estonia is created in Tallinn" (http://www.stat.ee/dokumendid/30210). Estonian Statistics Office. . Retrieved 2012-05-20.
[38] Regulation 1077/2011 establishing a European Agency for the operational management of large-scale IT systems in the area of freedom, security and justice (http://eur-lex.europa.eu/LexUriServ/LexUriServ.do?uri=OJ:L:2011:286:0001:0017:EN:PDF); http://ec.europa.eu/home-affairs/policies/agencies/agencies_intro_en.htm
[39] Ingrid Teesalu. "It's Official: Tallinn To Become EU's IT Headquarters" (http://news.err.ee/politics/50bcfc35-1061-48b9-b12a-e47594d7fc84). ERR. . Retrieved 27 April 2012.
[40] " Contact – AS Estonian Air (http://www.estonian-air.ee/estonian_air_contact)." Estonian Air. Retrieved on 18 January 2010.
[41] "Skype Jobs: Life at Skype". Jobs.skype.com.
[42] Steve Roman. "TeliaSonera Opens IT Development Center in Tallinn" (http://news.err.ee/economy/614ea892-6d60-4277-b18e-90acbc11a714). ERR. . Retrieved 7 June 2012.
[43] "Ericsson Eesti planning to invest EUR 6.4 mln > Tallinn" (http://www.tallinn.ee/eng/Ericsson-Eesti-planning-to-invest-EUR-6.4-mln). Tallinn.ee. . Retrieved 3 June 2011.
[44] Raivo Sormunen. "aripaev.ee – Skandinaavia uue börsifirma finantskeskus tuleb Tall" (http://www.ap3.ee/default.aspx?publicationid=ec804faa-77e0-4137-8169-ac76ac1d6758). Ap3.ee. . Retrieved 3 June 2011.
[45] "Tallinn, Estonia – Travel Photos by Galen R Frysinger, Sheboygan, Wisconsin" (http://www.galenfrysinger.com/estonia_tallinn.htm). Galenfrysinger.com. . Retrieved 3 June 2011.
[46] Copterline web page (http://www.copterline.com/page54_en.html)
[47] (http://www.copterline.com/offline/index.htm?target=Main&action=)
[48] "Twin Towns – Graz Online – English Version" (http://www.graz.at/cms/beitrag/10045157/606819/). graz.at. . Retrieved 5 January 2010.
[49] "*Groningen – Partner Cities*" (http://www.groningen.nl/functies/pagfunctie.cfm?parameter=1285). 2008 Gemeente Groningen, Kreupelstraat 1,9712 HW Groningen. . Retrieved 8 December 2008.
[50] "Malmö stads vänortssamarbete" (http://www.malmo.se/faktaommalmopolitik/internationelltsamarbete/vanortssamarbetet.4.33aee30d103b8f15916800032874.html) (in Swedish). 2004–2009 Malmö stad (http://www.malmo.se/), 205 80 Malmö, Organisationsnummer: 212000-1124. . Retrieved 27 June 2009.
[51] "Twin cities of Riga" (http://www.riga.lv/EN/Channels/Riga_Municipality/Twin_cities_of_Riga/default.htm). Riga City Council. . Retrieved 27 July 2009.
[52] "Vilniuse sõpruslinnad" (http://www.tallinn.ee/est/g1471s41613) (in Estonian). 2002–2009 Tallinn (http://www.tallinn.ee/). . Retrieved 12 September 2009.

External links

- The Website of the City of Tallinn (http://www.tallinn.ee/eng) (official)
- Panoramas of Tallinn Old Town (http://360.tallinn.ee/pano/list/en/)
- Tallinn on Wikitravel
- Tallinn (http://www.dmoz.org/Regional/Europe/Estonia/Harjumaa/Tallinn//) at the Open Directory Project

Helsinki

Helsinki
Helsinki – Helsingfors

— City —

Helsingin kaupunki
Helsingfors stad

From top-left: Helsinki Cathedral, Suomenlinna, Senate Square, Aurinkolahti beach, City Hall

Coat of arms

Nickname(s): Stadi, Hesa[1]

Helsinki

Location of Helsinki in Finland

Coordinates: 60°10′15″N 024°56′15″E

Country	Finland
Region	Uusimaa

Sub-region	Helsinki
Charter	1550
Capital city	1812
Government	
• **Mayor**	Jussi Pajunen
Area(2011-01-01)[2]	
• **City**	715.49 km^2 (**unknown operator: u'strong'** sq mi)
• **Land**	213.75 km^2 (**unknown operator: u'strong'** sq mi)
• **Water**	501.74 km^2 (**unknown operator: u'strong'** sq mi)
• **Urban**	770.26 km^2 (**unknown operator: u'strong'** sq mi)
• **Metro**	2970.18 km^2 (**unknown operator: u'strong'** sq mi)
Area rank	170 largest in Finland
Population (2012-01-31)[3]	
• **City**	596233
• **Rank**	Largest in Finland
• **Density**	2789.39/km^2 (**unknown operator: u'strong'**/sq mi)
• **Urban**	1060912
• **Urban density**	1377.34/km^2 (**unknown operator: u'strong'**/sq mi)
• **Metro**	1342469
• **Metro density**	451.98/km^2 (**unknown operator: u'strong'**/sq mi)
Population by native language[4]	
• **Finnish**	84.3% (official)
• **Swedish**	6.1% (official)
• **Others**	9.6%
Population by age[5]	
• **0 to 14**	13.7%
• **15 to 64**	72%
• **65 or older**	14.3%
Time zone	EET (UTC+2)
• **Summer (DST)**	EEST (UTC+3)
Municipal tax rate[6]	18.5%
Website	www.hel.fi [7]

Helsinki (listen; Swedish: ***Helsingfors***, listen) is the capital and largest city in Finland. It is in the region of Uusimaa, located in southern Finland, on the shore of the Gulf of Finland, an arm of the Baltic Sea. The population of the city of Helsinki is 596233 (31 January 2012),[3] making it by far the most populous municipality in Finland. Helsinki is located some 80 kilometres (**unknown operator: u'strong'** mi) north of Tallinn, Estonia, 400 kilometres (**unknown operator: u'strong'** mi) east of Stockholm, Sweden, and 300 kilometres (**unknown operator: u'strong'** mi) west of Saint Petersburg, Russia. Helsinki has close historical connections with these three cities.

The Helsinki Metropolitan Area includes the city of Helsinki and three other cities: Espoo and Vantaa, which immediately border Helsinki to the west and north, respectively; and Kauniainen, which is an enclave within the city of Espoo. The Helsinki metropolitan area is the world's northernmost urban area among those with a population of over one million people, and the city is the northernmost capital of a EU member state. Altogether 1.1 million people, approximately one in five Finns, live in the Helsinki Metropolitan Area.

Helsinki is Finland's major political, educational, financial, cultural and research centre as well as one of northern Europe's major cities. Approximately 70% of foreign companies operating in Finland have settled in the Helsinki region.[8]

The nearby city of Vantaa in the Helsinki metropolitan area is the location of Helsinki-Vantaa Airport, with frequent service to various destinations in Europe and Asia. Since early 2009, Helsinki has been exploring a merger with Vantaa. On 30 March 2009, the city council of Vantaa agreed to review Helsinki's proposal of a possible merger, while emphasizing that the review is not about the possibility of terminating the existence of the city of Vantaa.[9] On 31 January 2011, the city council of Vantaa turned down Helsinki's proposal of a possible merger, with 45 votes against the proposal compared to 22 in favour of it.[10]

In 2009, Helsinki was chosen to be the World Design Capital for 2012[11] by the International Council of Societies of Industrial Design, narrowly beating Eindhoven for the title.

History

Early history

Central Helsinki in 1820 before rebuilding. Drawing by Carl Ludvig Engel.

Helsinki was established as a trading town by King Gustav I of Sweden in 1550 as the town of Helsingfors, which he intended to be a rival to the Hanseatic city of Reval (today known as Tallinn). Little came of the plans as Helsinki remained a tiny town plagued by poverty, wars, and diseases. The plague of 1710 killed the greater part of the inhabitants of Helsinki.[12] The construction of the naval fortress Sveaborg (In Finnish *Viapori*, today also *Suomenlinna*) in the 18th century helped improve Helsinki's status, but it was not until Russia defeated Sweden in the Finnish War and annexed Finland as the autonomous Grand Duchy of Finland in 1809 that the town began to develop into a substantial city. During the war, Russians besieged the Sveaborg fortress and most of the city was devastated in a 1808 fire.

Czar Alexander I of Russia moved the Finnish capital from Turku to Helsinki in 1812 to reduce Swedish influence in Finland and bring the capital closer to St. Petersburg. Following the Great Fire of Turku in 1827, The Royal Academy of Turku, back then the country's only university, was also relocated to Helsinki, and eventually became the modern University of Helsinki. The move consolidated the city's new role and helped set it on the path of continuous growth. This transformation is highly apparent in the downtown core, which was rebuilt in neoclassical style to resemble St. Petersburg, mostly to a plan by the German-born architect C. L. Engel. As elsewhere, technological advancements such as railroads and industrialization were key factors behind the city's growth.

Twentieth century

Despite the tumultuousness of Finnish history during the first half of the 20th century, Helsinki continued its steady development. A landmark event was the XV Olympiad (1952 Olympic Games) held in Helsinki. Finland's rapid urbanization in the 1970s, occurring late relative to the rest of Europe, tripled the population in the metropolitan area, and the Helsinki Metro subway system was built. The relatively sparse population density of Helsinki and its peculiar structure have often been attributed to the lateness of its growth.

Name

The Swedish name *Helsingfors* (Swedish pronunciation: [hɛlsiŋˈfɔrs] or Swedish pronunciation: [hɛlsiŋˈfɔʂ]) is the original official name of the city of Helsinki (in the very beginning, spelled *Hellssingeforss*). The Finnish language form of the name probably originates from *Helsinga* and similar names used for the river that is currently known as the Vantaa River as documented already in the 14th century. *Helsinki* (Finnish pronunciation places stress on the first syllable: Finnish pronunciation: [ˈhelsiŋki]), is used to refer to the city in all languages except Swedish, and Norwegian. *Helsingfors* comes from the name of the surrounding parish, *Helsinge* (source for Finnish *Helsinki*) and the rapids (Swedish: *fors*), which flowed through the original village. The name *Helsinge* may have originated with medieval Swedish settlers who came from Hälsingland in Sweden. Others have proposed that the name derives from the Swedish word *hals* (neck), referring to the narrowest part of the river, i.e. the rapids.[13]

In Helsinki slang the city is nicknamed as either *Stadi* (from the Swedish word *stad*, meaning "city") or *Hesa* (short of Helsinki + slang suffix *sa*), with *Stadi* being used to assert that the speaker is native to the city.[1] *Helsset* is the Northern Saami name of Helsinki.

Geography

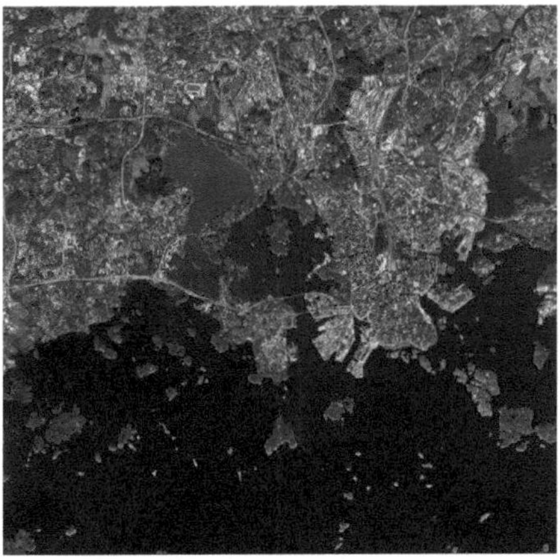

Parts of Helsinki and Espoo seen from Spot Satellite

Helsinki is spread across a number of bays and peninsulas and over a number of islands. The inner city area occupies a southern peninsula, which is rarely referred to by its actual name Vironniemi. Population density in certain parts of Helsinki's inner city area is very high, reaching 16494 inhabitants per square kilometre (**unknown operator: u'strong'** /sq mi) in the district of Kallio, but as a whole Helsinki's population density of 3050 per square kilometre (**unknown operator: u'strong'** /sq mi) ranks it as quite sparsely populated in comparison to other European capital cities. Much of Helsinki outside the inner city area consists of postwar suburbs separated from each other by patches of forest. A narrow, ten-kilometre-long (6 mi) Helsinki Central Park that stretches from the inner city to the northern border of Helsinki is an important recreational area for residents.

Major islands in Helsinki include Seurasaari, Lauttasaari and Korkeasaari – the latter being the site of the country's biggest zoo. Other significant islands are the fortress island of Suomenlinna (Sveaborg) and

Suomenlinna has been a UNESCO World Heritage Site since 1991.

the military island of Santahamina. Pihlajasaari island is a favourite summer spot for gay men and naturists, very much comparable to Fire Island off New York City.

Climate

Helsinki has a humid continental climate. Owing to the mitigating influence of the Baltic Sea and Gulf Stream, temperatures in winter are much higher than the far northern location might suggest, with the average in January and February around −5 °C (**unknown operator: u'strong'** °F).[14] Temperatures below occur normally a week or two in a year. However, because of the latitude, days last less than six hours around the winter solstice with really low sun rays, and the very cloudy weather at this time of year accentuates the darkness. Conversely, Helsinki enjoys long days in summer, close to nineteen hours around the summer solstice.[15] The average maximum temperature from June to August is around 19 to 21 °C (66 to 70 °F). Due to the marine effect, especially in the summer daily temperatures are cooler and night temperatures are higher than further away in the mainland. As near as in Vantaa, the climate is surprisingly much more continental, with warmer summers and colder winters. The highest temperature ever recorded at city centre was 34.0 °C (**unknown operator: u'strong'** °F) on 28 July 2010 and the lowest was −34.3 °C (−**unknown operator: u'strong'** °F) on 10 January 1987.[16]

Cityscape

Aleksanterinkatu in Central Helsinki.

Carl Ludvig Engel (1778–1840) was appointed to design a new city centre all on his own. He designed several neoclassical buildings in Helsinki. The focal point of Engel's city plan is the Senate Square. It is surrounded by the Government Palace (to the east), the main building of the University (to the west), and (to the north) the enormous Cathedral, which was finished in 1852, twelve years after C. L. Engel's death. Subsequently, Engel's neoclassical plan stimulated the epithet, *The White City Of The North.* Helsinki is, however, perhaps even more famous for its numerous Art Nouveau (Jugend in Finnish) influenced buildings of the romantic nationalism, designed in the early 1900s and strongly influenced by the Kalevala, which is a very popular theme in the national romantic art of that era. Helsinki's Art Nouveau style is also featured in large residential areas such as Katajanokka and Ullanlinna. The master of the Finnish Art Nouveau was Eliel Saarinen (1873–1950), whose architectural masterpiece was the Helsinki central railway station.

Helsinki also features several buildings by the world-renowned Finnish architect Alvar Aalto (1898–1976), recognized as one of the pioneers of architectural functionalism. However, some of his works, such as the headquarters of the paper company Stora Enso and the concert venue, Finlandia Hall, have been subject to divided opinions from the citizens.[17] [18] [19]

Aleksi in Central Helsinki.

Renowned functionalist buildings in Helsinki by other architects include the Olympic Stadium, the Tennis Palace, the Rowing Stadium, the Swimming Stadium, the Velodrome, the Glass Palace, the Exhibition Hall (now Töölö Sports Hall) and Helsinki-Malmi Airport. The sports venues were built to serve the 1940 Helsinki Olympic Games; the games were initially cancelled due to the Second World War, but the venues eventually got to fulfill their purpose in the 1952 Olympic Games. Many of them are listed by DoCoMoMo as significant examples of modern architecture. The Olympic Stadium and Helsinki-Malmi Airport are in addition catalogued by the National Board of Antiquities as cultural-historical environments of national significance.

The view across summertime Eläintarhanlahti.

As a historical footnote, Helsinki's neoclassical buildings were often used as a backdrop for scenes set to take place in the Soviet Union in many Cold War era Hollywood movies, when filming in the USSR was not possible. Some of the more notable ones are *The Kremlin Letter* (1970), *Reds* (1981) and *Gorky Park* (1983). Because some streetscapes were reminiscent of Leningrad's and Moscow's old buildings, they too were used in movie productions—much to some residents' dismay. At the same time the government secretly instructed Finnish officials not to extend assistance to such film projects.[20]

The Helsinki Cathedral is probably the most prominent building and symbol of the city.

A panoramic view over the southernmost districts of Helsinki from Hotel Torni. The Helsinki Old Church and its surrounding park are seen in the foreground, while the towers of St. John's Church (near center) and Mikael Agricola Church (right) can be seen in the middle distance, backdropped by the Gulf of Finland.

Parliament of Finland on the right, and new supplemental offices on the left.

A panoramic view of Helsinki Central Railway Station and its surroundings.

Cafe Regatta near Sibelius Park

A panoramic view of Kamppi Central and its surroundings.

Government

The city council of Helsinki consists of eighty-five members. The three largest parties are National Coalition (26), Greens (21), and Social Democrats (16).

Demographics

Helsinki has a higher proportion of women (53.4%) than elsewhere in Finland (51.1%). Helsinki's current population density of 2,739.36 people per square kilometer is by far the highest in Finland. Life expectancy for both genders is slightly below the national averages: 75.1 years for men as compared to 75.7 years, 81.7 years for women as compared to 82.5 years.[21] [22]

Helsinki has experienced strong growth since the 1810s, when it replaced Turku as the capital of the Grand Duchy of Finland, which later became the sovereign Republic of Finland. The city continued to show strong growth from that time onwards, with the exception during the Finnish Civil War period. From the end of World War II up until the 1970s there was a massive exodus of people from the countryside to the cities of Finland, in particular Helsinki. Between 1944 and 1969 the population of the city nearly doubled from 275,000[23] to 525,600.[24]

In the 1960s, the population growth of Helsinki proper began to ebb mainly due to lack of housing.[25] Many residents began to move to neighboring Espoo and Vantaa, where population growth has since soared. Espoo's population increased ninefold in sixty years, from 22,874 people in 1950 to 244,353 in 2009. Neighboring Vantaa has seen even more dramatic change in the same time span: from 14,976 in 1950 to 197,663 in 2009, a thirteenfold increase. These dramatic increases pushed the municipalities of greater Helsinki into more intense cooperation in such areas as public transportation[26] and waste management.[27] The increasing scarcity of housing and the higher costs of living in the Helsinki metropolitan area have pushed many daily commuters to find housing in formerly very rural areas, and even further, to such cities as Lohja (50 kilometres or 30 miles northwest from the city centre), Hämeenlinna and Lahti (both 100 kilometres or 60 miles from Helsinki, and Porvoo (50 kilometres to the east).

Language

Finnish and Swedish are the official languages of the municipality of Helsinki. The majority, or 84.3%[28] of the population, speak Finnish as their native language. A minority, at 6.1%, speak Swedish. Around 9.6% of the population speak a native language other than Finnish or Swedish. Helsinki slang combines influences from both Finnish, Swedish and Russian languages. Finnish today is the common language of communication between Finnish speakers, Swedish speakers and speakers of other languages (new Finns) in day-to-day affairs in the public sphere between unknown persons. In case a speaker's knowledge of Finnish is not known, English is usually spoken. Swedish is commonly spoken in city or national agencies specifically aimed at Finland-Swedish speakers, such as the Social Services Department on Hämeentie or the Luckan Cultural centre in Kamppi. Knowledge of Finnish is also essential in business and is usually a basic requirement in the employment market.[29]

Finnish speakers surpassed Swedish speakers in 1890 to become the majority of the city's population.[30] At the time, Helsinki's population was 61,530.[31]

Immigration

Helsinki is the global gateway of Finland. The city has Finland's largest immigrant population in both absolute and relative terms. There are over 130 nationalities represented in Helsinki. The largest groups are from Russia, Estonia, Sweden, Somalia, Serbia, China, Iraq and Germany.

Foreign citizens make up 5.2% (29,200) of the population, while foreign born make up 7.9% (44,400). Of the resident foreigners 33% are EU nationals. In 2008, 55,245[32] residents spoke a native language other than Finnish, Swedish or one of the three Sami languages spoken in Finland. As previously mentioned, the largest group of residents with a non-Finnish background come from Estonia (5,900), Russia (5,633), Somalia (2,400), China (1,150) and Thailand (680).[33] Half of the immigrant population in Finland lives in greater Helsinki, and one third in the city of Helsinki.[34] 44% of the African population in Finland lives in Helsinki. In 2001, out of all the EU-15 capitals, only Lisbon had a smaller share of immigrants in their population. There are some challenges to integration, as the immigrant unemployment rate on average is 2.5 times higher than native Finns.[35]

Economy

The Helsinki metropolitan area generates approximately one third of Finland's GDP. GDP per capita is roughly 1.3 times the national average.[36]

Kamppi Center, a shopping and transportation complex in the Kamppi district in the centre of Helsinki.

The metropolitan area's gross value added per capita is 200% of the mean of 27 European metropolitan areas, equalling those of Stockholm or Paris. The gross value added annual growth has been around 4%.[37]

83 of the 100 largest Finnish companies are headquartered in Greater Helsinki. Two-thirds of the 200 highest-paid Finnish executives live in Greater Helsinki and 42% in Helsinki. The average income of the top 50 earners was 1.65 million euro.[38]

The tap water is of excellent quality and it is supplied by 120 km (**unknown operator: u'strong'** mi) long Päijänne Water Tunnel, one of the world's longest continuous rock tunnels. Bottled Helsinki tap water is even sold to countries such as Saudi Arabia.[39]

Education

Helsinki has 190 comprehensive schools, 41 upper secondary schools and 15 vocational institutes. Half of the 41 upper secondary schools are private or state-owned, the other half municipal. Higher level education is given in eight universities (see the section "Universities" below) and four polytechnics.

Main building of the University of Helsinki.

Universities

- University of Helsinki
- Aalto University (The Schools of Science are located in Espoo)
- Hanken School of Economics
- Academy of Fine Arts
- Sibelius Academy
- Theatre Academy
- National Defence University

Haaga-Helia University Of Applied Sciences is the largest business polytechnic in Finland.

Polytechnics

- Haaga-Helia University Of Applied Sciences
- Helsinki Metropolia University of Applied Sciences
- Arcada University of Applied Sciences

Helsinki is one of the co-location centres of the Knowledge and Innovation Community (Future information and communication society) of The European Institute of Innovation and Technology (EIT).[40]

The educational department takes part in Lifelong Learning Programme 2007–2013 in Finland.

Culture

Kiasma is a contemporary art museum located at the heart of Helsinki.

The biggest historical museum in Helsinki is the National Museum of Finland, which displays a vast historical collection from prehistoric times to the 21st century. The museum building itself, a national romantic style neomedieval castle, is a tourist attraction. Other major historical museum is the Helsinki City Museum, which introduces visitors to Helsinki's 500-year history. The University of Helsinki also has many significant museums, including the University Museum and the Natural History Museum.

The Finnish National Gallery consists of three museums: Ateneum Art Museum for classical Finnish art, Sinebrychoff Art Museum for classical European art, and Kiasma Art Museum for modern art. The old Ateneum, a neo-Renaissance palace from 19th century, is one of the city's major historical buildings. All three museum buildings are state-owned through Senate Properties.

Helsinki has three major theatres: The Finnish National Theatre, the Helsinki City Theatre, and the Finland Swedish Svenska Teatern. The city's main musical venues are the Finnish National Opera, Helsinki Music Centre and the Finlandia concert-hall. Bigger concerts and events are usually held at one of the city's two big ice hockey arenas: the Hartwall Areena or the Helsinki Ice Hall. Helsinki has Finland's largest fairgrounds.

Many widely renowned and acclaimed bands have originated in Helsinki, including Hanoi Rocks, HIM, Stratovarius, The 69 Eyes, Norther, Wintersun, Finntroll, Ensiferum, The Rasmus and Apocalyptica.

Art

- The Helsinki Festival is an annual arts and culture festival, which takes place every August (including the Night of the Arts).
- Vappu "Vappu" is an annual carnival for students and workers.

Helsinki Arena hosted the Eurovision Song Contest 2007, the first ever Eurovision Song Contest arranged in Finland, following Lordi's win in 2006.

At the Senate Square in September / October 2010, the largest open-air art exhibition ever in Finland took place: About 1.4 million people saw the international exhibition of *United Buddy Bears*.

Helsinki is the 2012 World Design Capital, in recognition of the use of design as an effective tool for social, cultural and economic development in the city. In choosing Helsinki, the World Design Capital selection jury highlighted Helsinki's use of 'Embedded Design', which has tied design in the city to innovation, "creating global brands, such as Nokia, Kone and Marimekko, popular events, like the annual Helsinki Design Week, outstanding education and research institutions, such as the University of Art and Design Helsinki, and exemplary architects and designers such as Eliel Saarinen and Alvar Aalto".[11]

Media

National media

Further information: Finland#Media and communications

Helsinki is the centre of the Finnish media, most national newspapers as well as the broadcasters are located there.

Local media

There are two public service local radio stations in Greater Helsinki, Finnish-language *Ylen aikainen* (the regional edition of YLE Radio Suomi) and Swedish-language *YLE Radio Vega Huvudstadsregionen* (the regional edition of YLE Radio Vega). There are also privately owned local radio stations.

YLE produces local TV news programme for the region of Uusimaa, *Uudenmaan uutiset*.

Two daily newspapers, *Helsingin Sanomat* and *Hufvudstadsbladet* are published in Helsinki and function as the local as well as the national newspapers.

Sports

Helsinki has a long tradition of sports: the city gained much of its initial international recognition during the 1952 Summer Olympics, and the city has since then been very open to arranging sporting events, for example the first World Championships in Athletics 1983 and 2005, and European Championships in Athletics 1971, 1994 and 2012 etc. Helsinki hosts fairly successful local teams in both of the most popular team sports in Finland, football, ice hockey and synchronized skating. Helsinki houses Helsingin Jalkapalloklubi (HJK), Finland's largest and most successful football club. Helsinki's track and field club Helsingin Kisa-Veikot is also pretty dominant in Finland. Ice Hockey is a sport of passion for many Helsinki residents, who usually take a stance for either of the local clubs IFK Helsingfors (HIFK) or Jokerit. HIFK, with 14 Finnish championships titles, also plays in the highest bandy division,[41] so does Botnia –69. The Olympic stadium hosted the 1st ever Bandy World Championships in 1957.[42]

The Helsinki Olympic Stadium was the centre of activities during the 1952 Summer Olympics.

Transportation

Roads

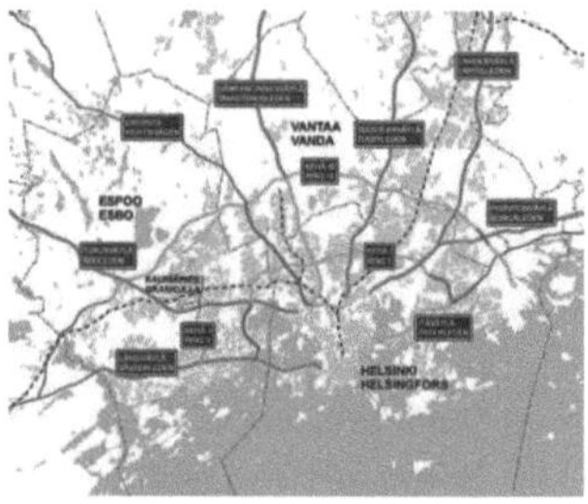

Helsinki region roads.

The backbone of Helsinki's motorway network consists of three semicircular ring roads, Ring I, Ring II, and Ring III, which connect together expressways heading to other parts of Finland, and the western and eastern arteries of *Länsiväylä* and *Itäväylä* respectively. While variants of a *Keskustatunneli* tunnel under the city centre have been repeatedly proposed, as of 2011 the plan remains on the drawing board.

Helsinki has some 390 cars per 1000 inhabitants.[43] This is less than in cities of similar density, for instance, Brussels' 483 per 1000 and Stockholm's 401, and Oslo's 413.[44] [45]

The Helsinki Metro with its characteristic bright orange trains is the world's northernmost subway.

Rail transport and buses

Public transportation is generally a hotly debated subject in the local politics of Helsinki. In Helsinki metropolitan area, public transportation is managed under Helsinki Region Transport, the metropolitan area transportation authority. The diverse public transport system consists of trams, commuter rail, the subway, bus lines and two ferry lines.

Today, Helsinki is the only city in Finland to have trams or subway trains. There used to be two other cities in Finland with trams: Turku and Viipuri (Vyborg, now in Russia), but both have since abandoned trams. The Helsinki Metro, opened in the year 1982, is the only subway system in Finland. In 2006, the construction of the long debated extension of the subway system west into Espoo was approved, and serious debate about an eastern extension into Sipoo has taken place.[46]

The possibility of a Helsinki to Tallinn Tunnel is currently being researched. The rail tunnel would connect Helsinki to the Estonian capital Tallinn, further linking Helsinki to the rest of continental Europe by Rail Baltica.

Malmi airport, one of the oldest in the world and Finland's- main general aviation airport.

Aviation

Air traffic is handled primarily from the international Helsinki Airport, located approximately 19 kilometres (**unknown operator: u'strong'** mi) north of Helsinki's downtown area, in the neighbouring city of Vantaa. Helsinki's second airport, Malmi Airport, is mainly used for general and private aviation. Helicopter flights to Tallinn are available from Hernesaari Heliport.

Sea transport

Ferry connections to Tallinn, Mariehamn and Stockholm are serviced by various companies. Finnlines passenger-freight ferries to Gdynia, Poland, Travemünde, Germany and Rostock, Germany are also available. St Peter Line offers passenger ferry service to Saint Petersburg several times a week.

Gallery

Skyline of central Helsinki, as seen from the roof of the Erottaja fire station.

Helsinki Central railway station

South harbour of Helsinki

The National Museum of Finland is located in Helsinki.

The Senaatintori square on a winter morning in December.

The Pohjoisranta at night.

Stockmann department store along the Aleksanterinkatu's Christmas street.

Restaurant Vltava on the right and Sanomatalo in the background.

Helsinki market square in winter.

Central Helsinki in evening.

Helsinki Olympic Stadium Tower, offers a good view over Helsinki.

Rooftops of the southern inner city districts.

The Uspenski Orthodox cathedral

Wall of Parliament House

The Esplanadi Park in central Helsinki in early june.

Finnish National Theatre and Aleksis Kivi statue in Rautatientori.

Altar of Temppeliaukio Church that is built underground

Hietaniemi beach in summer.

Itis shopping centre is the biggest shopping mall in the Nordic countries.

References

Notes

[1] Ainiala, Terhi (2009). "Place Names in the Construction of Social Identities: The Uses of Names of Helsinki" (http://hdl.handle.net/10315/2924). Research Institute for the Languages of Finland. . Retrieved 22 September 2011.

[2] "Area by municipality as of 1 January 2011" (http://www.maanmittauslaitos.fi/sites/default/files/pinta-alat_2011_kunnannimenmukaan.xls) (in Finnish and Swedish) (PDF). Land Survey of Finland. . Retrieved 9 March 2011.

[3] "Population by municipality as of 31 January 2012" (http://vrk.fi/default.aspx?docid=5919&site=3&id=0) (in Finnish and Swedish). *Population Information System*. Population Register Center of Finland. . Retrieved 16 February 2012.

[4] "Population according to language and the number of foreigners and land area km2 by area as of 31 December 2008" (http://pxweb2.stat.fi/Dialog/varval.asp?ma=060_vaerak_tau_107_fi&ti=Väestö+kielen+mukaan+sekä+ulkomaan+kansalaisten+määrä+ja+maa-pinta-ala+alueittain++1980+-+2008&path=../Database/StatFin/vrm/vaerak/&lang=3&multilang=fi). *Statistics Finland's PX-Web databases*. Statistics Finland. . Retrieved 29 March 2009.

[5] "Population according to age and gender by area as of 31 December 2008" (http://pxweb2.stat.fi/Dialog/varval.asp?ma=050_vaerak_tau_104_fi&ti=Väestö+iän+(1-v.)+ja+sukupuolen+mukaan+alueittain+1980+-+2008&path=../Database/StatFin/vrm/vaerak/&lang=3&multilang=fi). *Statistics Finland's PX-Web databases*. Statistics Finland. . Retrieved 28 April 2009.

[6] "List of municipal and parish tax rates in 2011" (http://www.vero.fi/nc/doc/download.asp?id=7996;193801). Tax Administration of Finland. 29 November 2010. . Retrieved 13 March 2011.
[7] http://www.hel.fi/
[8] Helsinki region in brief (http://web.archive.org/web/20071202040416/http://www.helsinkiregion.com/helsinki_region_in_brief/)
[9] Vantaa city council agrees to a review of a possible merger proposal (http://www.hs.fi/kaupunki/artikkeli/Vantaan+kaupunginhallitus+taipui+liitosselvitykseen/1135244782232), HS.fi. Retrieved 30 March 2009.
[10] Helsingin Sanomat, 1 February 2011, pp. A12-A13.
[11] "City of Helsinki appointed World Design Capital 2012" (http://www.worlddesigncapital.com/news/releases/main71.htm). Worlddesigncapital.com. 2009-11-25. . Retrieved 2010-04-13.
[12] "Ruttopuisto – Plague Park" (http://www.tabblo.com/studio/stories/view/409531/). Tabblo.com. . Retrieved 2008-11-03.
[13] "Utbildning & Vetenskap: Svenskfinland" (http://veta.yle.fi/svenskfinland/artikel.php?id=23&subject=mellannyland). Veta.yle.fi. . Retrieved 2009-07-08.
[14] "Climatological statistics for the normal period 1971–2000" (http://www.fmi.fi/weather/climate_6.html). Fmi.fi. . Retrieved 2010-04-13.
[15] Tukiainen, Matti. "Helsinki, Finland – Sunrise, sunset, dawn and dusk times around the World!" (http://www.gaisma.com/en/location/helsinki.html). *Gaisma*. . Retrieved 11 February 2011.
[16] "Ilmatieteen laitos – Sää ja ilmasto – Ilmastotilastot – Ilman lämpötila" (http://www.fmi.fi/saa/tilastot_4.html) (in Finnish). Finnish Meteorological Institute. . Retrieved 2009-07-08.
[17] "Stora Enson pääkonttori, Kanavaranta 1" (http://helsinginaallot.blogspot.com/2007/02/stora-enson-pkonttori-kanavaranta-1.html) (in Finnish). *"Helsingin Aallot" blog*. 25 February 2007. . Retrieved 5 February 2011.
[18] "Kohtaako Enson konttori voittajansa?" (http://www.hs.fi/paakirjoitus/artikkeli/Kohtaako+Enson+konttori+voittajansa/HS20080614SI1MA01re3) (in Finnish). *Helsingin Sanomat*: Lead editorial. 14 June 2008. . Retrieved 5 February 2011.
[19] Penttilä, Vappu. "Kiasma nousi inhokkien ykköseksi" (http://www2.hs.fi/extrat/kulttuuri/rakennukset/rakennukset_c2.html) (in Finnish). *Verkkoliite*. Helsingin Sanomat. . Retrieved 5 February 2011.
[20] Ministry for Foreign Affairs of Finland, Political Department: Memo 56 of 20 January 1982 (labelled highly confidential in 1982) (http://img.mtv3.fi/mn_liitteet/mtv3/uutiset/bulleting/300538.pdf)PDF (1.37 MB)
[21] Tapani Valkonen ym. (2007-12-17). "Tutkimuksia 10/2007: Elinajanodotteen kehitys Helsingissä ja sen väestönosaryhmissä 1991–2005" (http://www.hel2.fi/tietokeskus/julkaisut/pdf/07_12_17_tutk_10_Valkonen.pdf). . Retrieved 2007-12-30.
[22] Tilastolaitoksen historiaa. "Tilasto" (http://www.stat.fi/org/historia/elinajanodote.html). Stat.fi. . Retrieved 2010-04-13.
[23] "Helsingin historia" (http://www.hel.fi/wps/portal/Helsinki/Artikkeli?WCM_GLOBAL_CONTEXT=/helsinki/fi/Helsinki-tietoa+ja+linkkej_/Helsingin+historia). Hel.fi. . Retrieved 2010-04-13.
[24] "Maan alle" (http://www.aatos.fi/Hki450v/metro.html). Aatos.fi. 1972-12-30. . Retrieved 2010-04-13.
[25] Butzin, Bernhard (1991). "Helsinki — aspects of urban development and planning". *GeoJournal* (Springer, Netherlands) **2** (1): 11–26. doi:10.1007/BF00212573. ISSN 0343-2521.
[26] "HSL Helsingin seudun liikenne – About HSL" (http://www.hsl.fi/EN/abouthsl/Pages/default.aspx). Hsl.fi. 2010-01-01. . Retrieved 2010-04-13.
[27] "HSY – Default" (http://www.hsy.fi/en/Pages/default.aspx). Hsy.fi. . Retrieved 2010-04-13.
[28] "Taulukko: Väestö kielen mukaan sekä ulkomaan kansalaisten määrä ja maa-pinta-ala alueittain 1980 – 2008" (http://pxweb2.stat.fi/Dialog/Varvalagg.asp?ma=060_vaerak_tau_107_fi&ti=Väestö+kielen+mukaan+sekä+ulkomaan+kansalaisten+määrä+ja+maa-pinta-ala+alueittain++1980+-+2008&path=../Database/StatFin/vrm/vaerak/&lang=3&xu=&yp=&nr=1&aggfile(1)=Maakunnat+ja+kunnat+2009&prevagg=NNN&mapname=&multilang=fi). Pxweb2.stat.fi. 2001-08-13. . Retrieved 2010-04-13.
[29] "Immigrants Learning Swedish over Finnish Run into Problems | News | YLE Uutiset" (http://yle.fi/uutiset/teksti/news/2010/11/immigrants_learning_swedish_over_finnish_run_into_problems_2114684.html). yle.fi. . Retrieved 2011-09-16.
[30] "Helsingin nimistön vaiheita" (http://scripta.kotus.fi/www/verkkojulkaisut/julk125/helsinki/). Scripta.kotus.fi. . Retrieved 2010-04-13.
[31] IGS.kirjastot.fi (http://igs.kirjastot.fi/iGS/kysymykset/haku.aspx?word=Viipuri)
[32] "Tilasto" (http://pxweb2.stat.fi/database/StatFin/vrm/vaerak/vaerak_fi.asp). Pxweb2.stat.fi. . Retrieved 2010-04-13.
[33] Helsingin Sanomat (2008-12-07). "Helsingissä asuu virolaisia enemmän kuin venäläisiä" (http://www.hs.fi/kaupunki/artikkeli/HelsingissÃ¤+asuu+virolaisia+enemmÃ¤n+kuin+venÃ¤lÃ¤isiÃ¤/1135241719777). Hs.fi. . Retrieved 2010-04-13.
[34] Katriina Pajari (2008-12-07). "Kolmannes maahanmuuttajista asuu Helsingissä – HS.fi – Kaupunki" (http://www.hs.fi/kaupunki/artikkeli/Kolmannes+maahanmuuttajista+asuu+HelsingissÃ¤/1135241719769). HS.fi. . Retrieved 2010-04-13.
[35] Elsa Tuppurainen (2008-12-03). "Jussi Pajunen jarruttaisi maahanmuuttoa" (http://www.hs.fi/kaupunki/artikkeli/Jussi+Pajunen+jarruttaisi+maahanmuuttoa/1135241571625). Hs.fi. . Retrieved 2010-04-13.
[36] Helsinki Region Statistics (http://www.helsinginseutu.fi/hki/hs/The+Region+of+Helsinki/City+information+and+statistics/Helsinki+Region+Statistics) Retrieved 2010-12-10.
[37] "Kansilehti2.vp" (http://www.hel2.fi/Tietokeskus/julkaisut/pdf/04_11_09_suokas_vj31.pdf) (PDF). . Retrieved 2010-04-13.
[38] "Helsingin Sanomat – International Edition – Metro" (http://www.hs.fi/english/article/Head+offices+of+large+companies+bring+good+taxpayers+to+Helsinki+region/1101981569833). Hs.fi. 2005-11-09. . Retrieved 2009-07-08.
[39] Bottled water sells (http://www.finnfacts.com/english/main/actualities/pullovesi.htm), Finnfacts
[40] "European Institute of Innovation and Technology: Home" (http://eit.europa.eu/home.html). Eit.europa.eu. . Retrieved 2010-04-13.

[41] Video from the Finnish final 2009 against OLS from Oulu: Youtube.com (http://www.youtube.com/watch?v=Hn-j0t5yxIE&feature=related)
[42] Video from the tournament (http://www.yle.fi/player/player.jsp?name=Elävä+arkisto/02869_1)
[43] This is Helsinki (http://www.hel2.fi/tietoa/helbro1.pdf), by City of Helsinki
[44] EAUE.de (http://www.eaue.de/Promode/Runge.pdf)
[45] "Tietokeskus: suunnatframe" (http://www.hel2.fi/tietokeskus/suunnat/ss405/suunnatframe2.html). Hel2.fi. . Retrieved 2009-07-08.
[46] Lansimetro.fi (http://www.lansimetro.fi/) – an information portal dedicated to the "länsimetro" subway expansion in the Helsinki capital region.

External links

- Helsinki travel guide from Wikitravel
- Scenaries, sights, and places to go in and near Helsinki in Wikimedia Commons
- Hel.fi – Official website (http://www.hel.fi/wps/portal/Helsinki_en/?WCM_GLOBAL_CONTEXT=/en/Helsinki/) (in English; other languages available)
- Visithelsinki.fi – Official City of Helsinki Tourism Website (http://www.visithelsinki.fi/)
- Helsinki.mobi – Official mobile website (http://helsinki.mobi/)
- Helsinki.fi – City Portal (http://www.helsinki.fi/en/index.shtml)
- World Design Capital 2012 Helsinki (http://www.wdc2012helsinki.fi/en) Official site

Rostock

Rostock

New Market (Neuer Markt) in Rostock

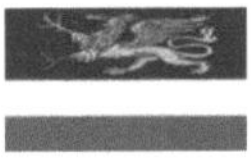

Rostock

Coordinates	54°5′0″N 12°8′0″E
Administration	
Country	Germany
State	Mecklenburg-Vorpommern
District	Urban district
City subdivisions	21 boroughs
Lord Mayor	Roland Methling (Ind.)
Basic statistics	
Area	181.44 km^2 (**unknown operator: u'strong'** sq mi)
Elevation	13 m (43 ft)
Population	200413 *(31 December 2007)*[1]

- Density	1105 /km^2 (2861 /sq mi)
Other information	
Time zone	CET/CEST (UTC+1/+2)
Licence plate	HRO
Postal codes	18001–18147
Area code	0381
Website	www.rostock.de [2]

Rostock (German pronunciation: [ˈʁɔstɔk]) is the largest city in the north German state Mecklenburg-Vorpommern. Rostock is located on the Warnow river; the quarter of Warnemünde 12 km north of the city centre lies directly on the coast of the Baltic Sea.

The city territory of Rostock stretches for about 20 kilometres along the Warnow to the Baltic Sea. The largest built-up area of Rostock is on the western side of the river. The eastern part of its territory is dominated by industrial estates and the forested area of the Rostock Heath.

History

Early history

Confirmation of Lübeck law city rights, 1218

In the 11th century Polabian Slavs founded a settlement at the Warnow river called *Roztoc* (which means *broadening of a river*); the name Rostock is derived from that designation. The Danish king Valdemar I set the town aflame in 1161.

Afterwards the place was settled by German traders. Initially there were three separate cities:

1. *Altstadt* (Old Town) around the *Alter Markt* (Old Market) with St. Petri (St. Peter's Church),
2. *Mittelstadt* (Middle Town) around the *Neuer Markt* (New Market) with St. Marien (St. Mary's Church) and
3. *Neustadt* (New Town) around the *Hopfenmarkt* (Hop Market, now University Square) with St. Jakobi (St. James's Church, now demolished).

Hanseatic League

The rise of the city began with its membership of the Hanseatic League. In the 14th century it was a powerful seaport town with 12,000 inhabitants and the biggest city of Mecklenburg. Ships for cruising the Baltic Sea were constructed in Rostock. In 1419 one of the oldest universities in Northern Europe, the University of Rostock, was founded.

15th to 18th century

At the end of the 15th century the dukes of Mecklenburg succeeded in enforcing their rule over the town of Rostock, which had until then been only nominally subject to their rule and essentially independent. They took advantage of a riot known as *Domfehde*, a failed uprising of the impoverished population. Subsequent quarrels with the dukes and persistent plundering led ultimately to a loss of economic and political power.

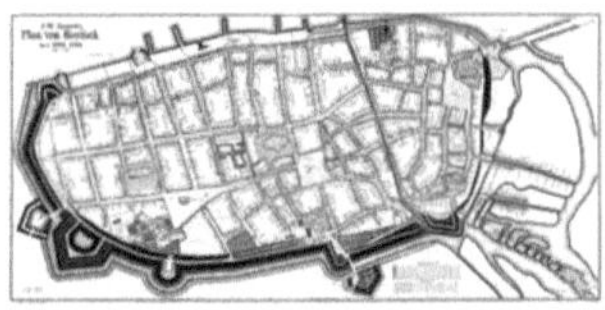

Rostock 1780-90

In 1565 there were further clashes with Schwerin that which had far-reaching consequences. Among other things, was the introduction of a beer excise that favoured the dukes. John Albert I advanced on the city with 500 horsemen, after Rostock had refused to take the formal oath of allegiance, and had the city wall slighted in order to have a fortress built. Not until the first Rostock Inheritance Agreement of 21 September 1573, in which the state princes were guaranteed hereditary rule over the city for centuries, binding Rostock for a long time, and recognizing them as the supreme judicial authority, was the conflict ended. The citizens slighted the fortress the following spring. From 1575 to 1577 it the city walls were rebuilt, as was the *Lagebusch* tower and the *Stein* Gate in the Dutch Renaissance style. The inscription *sit intra te concordia et publica felicitas*, which can still be read on the gate, refers directly to the conflict with the Duke. In 1584 it finally came to the Second Rostock Inheritance Agreement, which resulted in a further loss of former tax privileges. At the same time, these inheritance contracts put paid to Rostock's ambition of achieving imperial immediacy as Lübeck had done in 1226.

The strategic location of Rostock provoked the envy of its rivals. Danes and Swedes occupied the city twice, first during the Thirty Years' War (1618–48) and again from 1700 to 1721. Later, the French, under Napoleon, occupied the town for about a decade until 1813. It was in nearby Lübeck-Ratekau that Blücher, who was actually born in Rostock and who was one of few generals to fight on after the Battle of Jena, surrendered to the French in 1806. This was only after furious street fighting in the Battle of Lübeck, in which he led some of the cavalry charges himself; the exhausted Prussians had, by the time of the surrender, neither food nor ammunition.

19th century

In the first half of the 19th century Rostock regained much of its economic importance, at first due to the wheat trade, and, from the 1850s, to industry, especially to its shipyards. The first propeller-driven steamers in Germany were constructed here.

The city grew in size and population, with new quarters emerging in the south and west of the ancient borders of the city. Two notable developments were added to house the increasing population at around 1900:

1. *Steintor-Vorstadt* in the south, stretching from the old city wall to the facilities of the new *Lloydbahnhof* Railway Station (now Hauptbahnhof). It was designed as a living quarter and consists mostly of large single houses, once inhabited by wealthy citizens.
2. *Kröpeliner-Tor-Vorstadt* in the west, designed to house the working population as well as smaller and larger industrial facilities such as *Mahn & Ohlerich's Brewery* (now Hanseatische Brauerei Rostock). The main shipyard, *Neptun* was just nearby at the shore of the river.

20th century

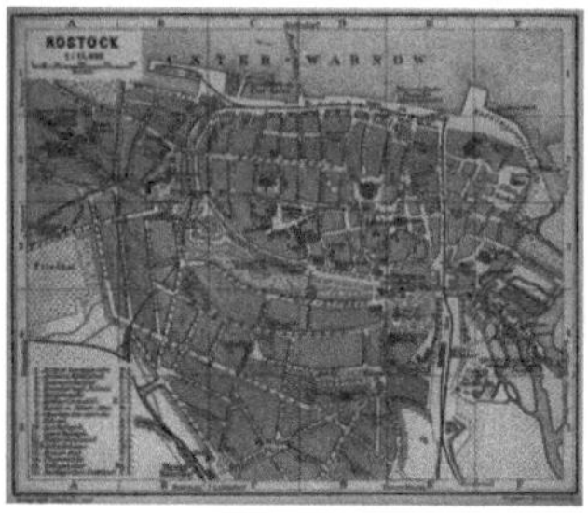

Rostock 1910

In the 20th century, important aircraft manufacturing facilities were situated in the city, such as the Arado Flugzeugwerke in Warnemünde and the Heinkel Works with facilities at various places. It was at their facilities in Rostock-Marienehe that the world's first jet plane made its test flights. Aeroplane construction ceased at the end of the Second World War.

Large parts of the central city were destroyed in World War II by Allied bombing in 1942 and 1944. Through reconstruction and subsequent extension, the city became a major industrial centre of the German Democratic Republic with the port being developed as the primary gate to the world. Much of the historic centre has been faithfully rebuilt and much of its historic character restored. This includes several buildings characterised by vertical brick ribs, a style common to the Hanseatic towns.

The destroyed city in 1942

Following the reunification of Germany in 1990, Rostock lost its prior privileged position as the principal overseas port of the former GDR and became one of several German ports, now located in one of the least industrialised regions of reunited Germany. Despite large infrastructure investments, the city's economy declined in the 1990s but is now growing again.

Rostock's population dropped from nearly 260,000 in 1989 to about 200,000 today, primarily due to suburbanisation but also due to emigration to more prosperous western regions of Germany.

Politics

Symbols

Coat of Arms

In Rostock's long history, the city carried three different coats of arms known as the *Signum*, *Secretum* and *Sigillum*. The Signum, which can be traced back to 1367, was developed last and is to this day the coat of arms of the city.

The flag depicts a golden griffin on a blue background as well as the colors of the Hanseatic League, silver and red.

The coat of arms can not only be seen on flags, houses and at bus stops, but also on bridges, gullies, fences, ships and restaurants.

Administration

Since the 13th century, the governing body of the city has been the city council (*Rat*), first consisting of ten, later of 24 aldermen (*Ratsherren*). The chairman of the city council was the city mayor. In the 19th century there were even three mayors. Since 1925, the head of the city has born the title of Lord Mayor. Having been elected for centuries by the city council, he is now elected directly by the citizens of Rostock, following a reform in 2002.

SIT INTRA TE CONCORDIA ET PUBLICA FELICITAS

Motto: Within your walls may it be harmony and happiness

The city parliament (*Bürgerschaft*) represents the citizens. Representative are elected for five years. The number of representatives is currently 53.

- 13 Linke
- 10 SPD
- 9 CDU
- 5 Greens
- 4 FDP
- 4 FÜR ROSTOCK - pro OB
- 3 Rostocker Bund
- 5 others

City Hall

The city parliament is presided by the *Präsident der Bürgerschaft*. He heads and prepares the sessions and, together with the Lord Mayor, represents the city.

Roland Methling (Independent), was elected Lord Mayor of Rostock in the first round by 58,2% of the voters on 27 February 2005.

Partner cities

Rostock has signed partnership agreements with the following cities:

- Szczecin, Poland, since 1957
- Turku, Finland, since 1959
- Dunkirk, France, since 1960
- Riga, Latvia, since 1961[3]
- Antwerpen, Belgium, since 1963
- Århus, Denmark, since 1964
- Gothenburg, Sweden, since 1965
- Bergen, Norway, since 1965
- Varna, Bulgaria, since 1966
- Rijeka, Croatia, since 1966
- Bremen, Germany since 1987
- Dalian, People's Republic of China since 1988
- Raleigh, USA, since 2001

Moreover, Rostock is a member of the international network New Hanse.

Regiopolis Rostock

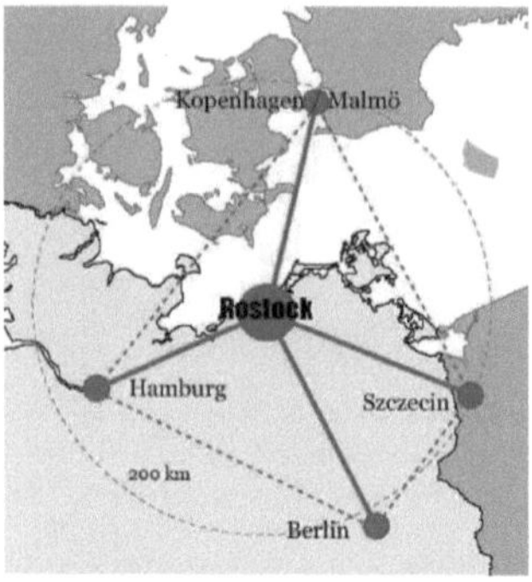

Geographical position of the Rostock Regiopolis

Rostock is the first German city region that defines itself not only as a city, but as a regiopolis. A regiopolis can be compared with a metropolis, but on a smaller scale. This is a sign for the intraregional cooperation that can be found within the region. A taskforce with different actors such as the hanseatic city of Rostock, the administrative district of Rostock, the Regional Planning Association Mid Mecklenburg/Rostock and the local business organizations is working on the promotion of the concept.[4]

Geography

Rostock is located nearly centrally on Mecklenburg-Vorpommern's Baltic Sea coast. The city is crossed by the Warnow.

The seaside part of Rostock, Rostock-Warnemünde, is about 16 km to the north of the historic city centre. The west and the southeast are the most densely populated parts of town. The overseas port is to the east of Rostock. Rostock stretches 21.6 kilometres from the Baltic Sea to the south and 19.4 km from east to west.

Climate

As with most of Germany,Rostock has cold winters,cool summers and low sunshine levels.

Main sights

Rostock

Panorama of Rostock from the bank of the Warnow river during the Hanse Sail

Heiligengeisthof (Holy Spirit Courtyard).

One of the most picturesque places in Rostock is the *Neuer Markt* (New Market Square), with the Town Hall (originally built in the 13th century in Brick Gothic style, but extensively transformed in the 18th century, with the addition of a Baroque facade and a Banqueting Hall. The square also preserved six original, beautifully restored, gable houses from the 15th and 16th centuries. (The rest of the old houses in Hanseatic style that once bordered the square were destroyed in an Allied air-raid in 1942.)

The 15th-century *Kerkhofhaus* (at Große Wasserstraße, behind the Town Hall) is considered the best preserved brick Gothic house in Rostock.

St. Mary`s Church *Marienkirche*, on Ziegenmarkt, is an imposing Brick Gothic church. Built in the 13th century, it was enlarged and modified at the end of the 14th century into the present cross-shaped basilica. The huge tower was not completed until the end of the 18th century. Inside there is an astronomical clock built in 1472 by Hans Düringer.

Kröpeliner Straße - the main shopping street.

The main pedestrian precinct is *Kröpeliner Straße*, that runs east from the Neuer Markt to the 14th-century *Kröpeliner Tor*, a former town gate. The main buildings of Rostock University lie at Universitätsplatz, near the middle of the street, in front of the lively fountain of *zest for life (Brunnen der Lebensfreude).*

The *Kloster St Katharinen* (Convent of St. Catherine), an old Franciscan monastery founded in 1243, and extended several times during the 14th and 15th centuries. Now used as the seat of the Academy of Music and Theatre (HMT-Rostock) [6].

The Brick Gothic *Nikolaikirche* (St. Nicholas Church), which is the oldest church in Rostock, built in mid-13th century. Heavily damaged during World War II and subsequently restored, the building is now used as an exhibition centre and concert hall, due to its outstanding acoustics.

Some parts of the medieval city wall, with four city gates, have survived to the present day.

Warnemünde

Warnemünde is the seaside part of Rostock and a major attraction of the city. Locals and tourists alike enjoy the maritime flair of old houses, a large beach, a lighthouse and the old fisherman port.

Alexandrinenstraße in Warnemünde.

Economy

The economy is strongly influenced by tourism, the University of Rostock and maritime industries (especially shipbuilding) and the service sector. Major companies include:

- Aker Warnow Werft, shipyard belonging to Aker Yards
- Neptun Werft, shipyard belonging to Meyer Neptun Group
- Deutsche Seereederei Rostock, transport, cruises, property and tourism holding
- Suzlon,worlds 5th largest wind turbine manufacturers
- Nordex, a major producer of wind turbines
- Hanseatische Brauerei Rostock, German brewery belonging to the Oetker-Gruppe
- Liebherr, manufacturer of cranes
- Caterpillar Inc., manufacturer of diesel engines for ships
- IKEA
- Scandlines, German-Danish ferry operator
- Inros Lackner AG, Architects, Engineers, Consulting company
- Yara International, supplier of plant nutrients
- AIDA Cruises, German company for cruises

Speicher (office buildings) at night. Headquarters of AIDA Cruises.

Furthermore, Rostock is the seventh-largest port of the Baltic Sea, and among the largest in Germany.

Education

Rostock is home to one of the oldest universities in the world. Founded in 1419, the University of Rostock is the third oldest university in Germany in continuous operation, the second oldest in Northern Europe (after St Andrews) and the oldest university in continental northern Europe. It offers graduate and postgraduate programmes in evangelical theology, philosophy and arts, natural sciences and mathematics, law, engineering and naval architecture, agriculture and environmental science, medicine, state, and political and social science, and also maintains a botanical garden (the Botanischer Garten Universität Rostock).

University of Rostock.

The Academy of Music and Theatre [7] (*Hochschule für Musik und Theater*) offers graduate degrees in artistic fields. Founded in 1994, the institution combined the former drama school *Ernst Busch* and the outpost school of the Hanns Eisler Music School Berlin. Today, the school is a member of the Association of Baltic Academies of Music (ABAM), a union of 17 music conservatories at the Baltic Sea and Israel. Unique in Europe is the postgraduate degree in piano duo performance. The school possesses a large opera stage (Katharinensaal) and two chamber music halls. There are concerts every day through the whole year.

Rostock hosts also the Max Planck Institute for Demographic Research, the Leibniz Institute for Catalysis as well as two branches of Fraunhofer Institutes, one for Computer Graphic and one for Large Structures in Production Technology.

Culture

Events

The city is home to the annual Hanse Sail festival, during which many large sailing ships and museum vessels are brought out to sea, drawing over 1.5 million visitors.

There is an annual jazz festival taking place in June called *Ostsee-Jazz* ("Baltic Sea Jazz").

Further events include:

Theater im Stadthafen

- Kurfilmfestival FiSh
- Rostocker Kulturwoche
- Sommer der Kulturen
- Rostocker Hafenfest
- Boulevardfest
- Warnemünder Woche

Museums and Zoo

- Rostock Art Gallery (*Kunsthalle Rostock*)
- Museum of Cultural History (*Kulturhistorisches Museum*)
- Stasi Museum (*Dokumentations- und Gedenkstätte der Bundesbeauftragten für die Unterlagen des Staatssicherheitsdienstes der ehemaligen Deutschen Demokratischen Republik*)
- Warnemünde Local History Museum (*Heimatmuseum Warnemünde*)
- Shipbuilding and Shipping Museum (*Schiffbau- und Schifffahrtsmuseum*)
- Rostock Zoo
- Walter Kempowski Archive

Walter Kempowski archives

Music and theatre

- Volkstheater Rostock
  - Norddeutsche Philharmonie
  - Rostocker Singakademie
- Niederdeutsche Bühne Rostock
- Compagnie de Comédie
- Kleine Komödie Warnemünde
- Mechaje
- Bühne 602
- Festspiele Mecklenburg-Vorpommern
- Ostsee Big Band (Jazz)

Sport

DKB-Arena, home ground of Hansa Rostock

Club	Sport	Founded	League	Venue	Head Coach
F.C. Hansa Rostock	Football	1965	2. Bundesliga	DKB-Arena	Wolfgang Wolf
HC Empor Rostock	Team handball	1954	2. Bundesliga	Rostocker Stadthalle	Maik Handschke
Piranhas Rostock	Ice hockey	1990	Oberliga (3rd division)	Eishalle Rostock	Henry Thom

Transport

Car

Rostock can be reached by motorway (*Autobahn*) A 1 from Hamburg via Lübeck on A 20 and by A 19 from Berlin and A 20 from Stettin in Poland.

Rostock Hauptbahnhof (Central Station)

The cruise ship A'Rosa Blu leaving Rostock in summer 2003

Public transport

Rostock Hauptbahnhof (Rostock Central Station) offers fast rail connections to Hamburg and Berlin and from there to almost any other European city.

Within the city a wide network of trams, buses and ferries is available. The first privately financed tunnel in Germany crosses the Warnow river and thus connects the eastern part of Rostock with the western part.

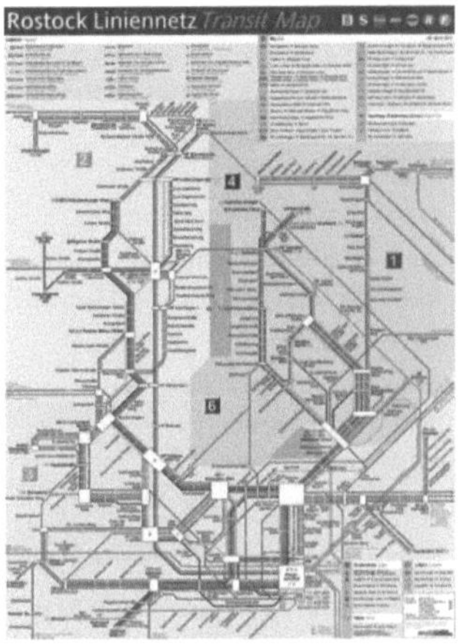

Transit map of Rostock (4 January 2012)

Ferry / Ship

Rostock is Germany's largest Baltic port. Rostock is also home to a large ferry port. It is a main base for ferry operators Scandlines, Tallink and TT-Line, which both connect Rostock with major Scandinavian destinations. Furthermore, Rostock receives the highest numbers of cruise tourists in Germany per year.

Ferries leave for

- Tallinn, Estonia
- Helsinki, Finland
- Gedser, Denmark
- Trelleborg, Sweden

Rostock harbour at sunset

Plane

The nearest international airports are in Hamburg and Berlin. There are connecting flights via Munich to Rostock-Laage Airport. There are also a number of airfields for smaller aircraft, e.g. Purkshof.

Notable people

This is an incomplete list of notable people who were born or lived in, or contributed to the welfare of, the City of Rostock:

- Gebhard Leberecht von Blücher, Prussian Generalfeldmarschall
- John Brinckman, writer
- Walter Kempowski, writer
- Ernst Heinkel, aviation pioneer
- Walter Hallstein, politician
- Joachim Gauck, president of Germany since 18 March 2012
- Jan Ullrich, cyclist
- Britta Kamrau, swimmer
- Paul Walden, scientist
- André Greipel, cyclist
- Paul Martens, cyclist
- *people who studied or did research in Rostock*:
  - Heinrich Schliemann, finished his PhD
  - Albert Einstein, received his first Doctor honoris causa in Rostock
  - Magnus Pegel, mathematician
  - Fritz Reuter, writer
  - Rudolf Steiner, founder of Anthroposophy
  - Uwe Johnson, writer
  - Arno Esch, politician

References

[1] Mecklenburg-Vorpommern Statistical Office. "Population of the districts *Ämter* and Municipalities, 30.06.2006" (http://www.statistik-mv.de/berichte/a__/a-i__/a123__/2006/daten/a123-2006-21.pdf) (in German). .
[2] http://www.rostock.de/
[3] "Twin cities of Riga" (http://www.riga.lv/EN/Channels/Riga_Municipality/Twin_cities_of_Riga/default.htm). Riga City Council. . Retrieved 2009-07-27.
[4] http://www.regiopole-rostock.de/
[5] "Average Conditions Rostock, Germany" (http://www.bbc.co.uk/weather/world/city_guides/results.shtml?tt=TT003630). BBC Weather. . Retrieved December 21, 2009.
[6] http://www.hmt-rostock.de/englisch/e_start.html
[7] http://www.hmt-rostock.de

External links

- Official website (http://http://www.rostock.de)
- at a glance (http://www.historicgermany.com/625.html)
- European Route of Brick Gothic: Rostock (http://www.eurob.org/index.php5/Hanseatic+city+of+Rostock;90/4)

MS *Star* approaching Helsinki, April 2007.

Career	
Name:	*Star*
Owner:	Tallink
Operator:	Tallink
Port of registry:	Tallinn, Estonia
Route:	Helsinki–Tallinn
Ordered:	1 August 2005
Builder:	Aker Finnyards, Helsinki, Finland
Yard number:	1356
Laid down:	30 May 2006
Launched:	23 November 2006
Christened:	23 November 2006
Acquired:	10 March 2007
In service:	12 April 2007
Identification:	IMO number: 9364722
Status:	In service
General characteristics	
Type:	Fast ropax ferry
Tonnage:	36,249 GT
Displacement:	3,500 metric tons deadweight (DWT)
Length:	186.00 m ()
Beam:	27.70 m ()
Draught:	6.50 m ()
Ice class:	1 A
Propulsion:	4 × MaK diesels combined 48000 kW (**unknown operator: u'strong'** hp)

Speed:	27 kn (**unknown operator: u'strong'** km/h; **unknown operator: u'strong'** mph)
Capacity:	1,900 passengers 520 passenger beds 450 cars 2,000 lanemeters

MS *Star* is a fast Ro-Pax ferry operated by the Estonian ferry company Tallink. She was built at Aker Finnyards Hietalahti shipyard, Helsinki, Finland and entered service on Tallink's Helsinki–Tallinn service on 12 April 2007.

Star was the first fast ferry capable of around-the-year service used on the Helsinki–Tallinn route.[1] The ship's design is heavily based on that of MS *SeaFrance Rodin*, built by Aker Finnyards for SeaFrance in 2001.[2] Originally it was reported the *Star* would have no passenger beds, but later specifications included beds for 520 passengers, meaning that the ship can offer short overnight cruises to Tallinn as well as a transportation service during the day. The green external livery of the ship is according to Tallink promotion material "meant to reflect the environmentally friendly aspects of the ship's design".[3]

Star has three daily departures from both Helsinki and Tallinn, with one crossing taking two hours. In Tallink marketing, *Star* has often been referred to as "Tallink Star". The ship's planned original delivery date was 5 April 2007, but this was postponed until 12 April 2007, owing to problems with the main engines.[4]

References

[1] Tallink press release: Tallink's New Vessel *Star* Started on Tallinn–Helsinki Route (http://www.tallink.com/mainMenu/pressRoom/pressReleases/pr120407.htm), retrieved 8. 10. 2007

[2] (Finnish) FCBS Forum: Tallink tilasi nopeakulkuisen lautan Aker Finnyardsilta (http://www.fcbsweb.com/forum/index.php?topic=9818.msg51217#msg51217), retrieved 8. 10. 2007

[3] Tallink press release: Tallink's New Vessel Named *Superstar* (http://www.tallink.com/mainMenu/pressRoom/pressReleases/pr071007.htm), retrieved 8. 10. 2007

[4] Turun Sanomat: M/S Star delayed by a week (http://www.turunsanomat.fi/talous/?ts=1,3:1004:0:0,4:4:0:1:2007-03-29,104:4:448635,1:0:0:0:0:0:), retrieved on 29. 3. 2007

External links

- Tallink official website for *Star* (http://www.tallinksilja.com/en/tab2/ships/star/)
- (Swedish) Fakta om Fartyg: M/S *Star* (2007) (http://www.faktaomfartyg.nu/star_2007.htm)
- Star at marinetraffic.com (http://www.marinetraffic.com/ais/shipdetails.aspx?MMSI=276672000)

Paldiski

Paldiski

Paldiski

Flag

Coat of arms

Location of Paldiski

Coordinates: 59°21′N 24°03′E

Country	Estonia
County	Harju County
City rights	1783[1]
Area	
• **Total**	34 km^2 (**unknown operator: u'strong'** sq mi)
Population (January 2010)	

• **Total**	4372
Time zone	EET (UTC+2)
• **Summer (DST)**	EEST (UTC+3)
Website	www.paldiski.ee [2]

Paldiski is a town and Baltic Sea port situated on the Pakri peninsula of north-western Estonia. Originally a Swedish settlement known as *Rågervik*, it became a Russian naval base in the 18th century. The Russians renamed it *Балтийский Порт* ("Baltiyskiy Port", i.e. Baltic Port, German: *Baltisch-Port*) in 1762. In Estonian the name was spelled *Baltiski* until 1933 when the Estonian pronunciation *Paldiski* became the official name.

History

Abandoned military buildings in Paldiski (1999).

Soviet era

In 1962, Paldiski became a Soviet Navy nuclear submarine training centre. With two land-based nuclear reactors, and employing some 16,000 people, it was the largest such facility in the Soviet Union. Because of its importance, the whole city was closed off with barbed wire until the last Russian warship left in August 1994. Russia finally relinquished control of the nuclear reactor facilities in September 1995.

Post-independence

After Estonia restored its independence, the city had not enough Estonian citizens and Paldiski was subordinated to Keila until 30 October 1996. Located some 45 km west of Tallinn, Paldiski is a municipality within Harju County and has a regular ferry connection with Kapellskär in Sweden (operated by Tallink). Derelict Soviet-style apartment buildings make up much of the town, and the relics of military bases are widespread. The majority of its residents are ethnic Russians.

Facilities today

Paldiski railway station in 2011.

Today the town has about 4300 inhabitants. Until recently, the amenities were limited to a single, small hotel (Valge Laev, with six guest bedrooms and a restaurant/bar) . Now there are also two grocery stores, an alcohol store, a bank, a tavern, and a bed & breakfast. The renovated train station also has a small cafe with hot food.

A great deal of investment has been put into the port facilities with a number of new berths having been created. Transfennica run a number of ships into the port from Hanko (Finland) and Lübeck (Germany).

Pakri wind farm is located in Paldiski at the tip of the Pakri peninsula near the old light house. It consists of 8 wind turbines of type Nordex N-90 and generates 18.4 MW of clean electricity.[3]

The old soviet 'pentagon' building has been demolished at some point between 2006 and 2009 to make way for a large and modern logistics park. [4].

The housing blocks in the town no longer appear dilapidated and abandoned and have been refitted and re-painted in recent years. There are also several new housing apartment buildings, plus the green areas and children's parks have been restored.

Town government building.

Housing in one of the poorer areas of the city.

Paldiski has a terminus station on the Elektriraudtee rail line, providing a convenient link to the capital city. As part of the town's redevelopment this once near-derelict station has been overhauled and painted in bright yellow and white colours.

Paldiski has two schools: Paldiski Gümnaasium [5] and Vene Gümnaasium.

There is one church, dedicated to St Nicholas, and part of the Estonian Evangelical Lutheran Church. Although closed for some years, this church has now reopened, with services held every Sunday lunchtime.[6] There is also a Pentecostal church, Methodist Church and Orthodox church.

The 2002 film *Lilya 4-ever*, set in an unnamed "former republic of the Soviet Union", was largely filmed in Paldiski.[7]

Notable people

- Amandus Adamson (1855–1929), Estonian sculptor and painter; lived and worked in Paldiski
- Salawat Yulayev (1754–1800), Bashkir national hero

References

[1] "About Paldiski" (http://www.paldiski.ee/index.php?id=11883). Paldiski Linn. . Retrieved 28 May 2012.
[2] http://www.paldiski.ee/
[3] "Pakri wind farm" (http://www.4energia.ee/en/projects/pakri-wind-park/). 4Energia. . Retrieved 28 May 2012.
[4] http://www.rrk.ee/?op=body&id=352
[5] http://www.paldgym.edu.ee/
[6] Service and contact details (http://www.eelk.ee/kontakt.php#paldiski) from the Estonian Lutheran Church's national website.
[7] Urbanik, Helen (2002-11-28). "Moodysson avastas Eestist head vene näitlejad" (http://paber.ekspress.ee/viewdoc/60CCAFAD6A66AEB0C2256C7C0027D7E5) (in Estonian). *Eesti Ekspress*. . Retrieved 29 August 2009.

External links

- Population of Paldiski by mother tongue in 1897 (http://demoscope.ru/weekly/ssp/rus_lan_97_uezd_eng.php?reg=1732)

Stena_Line

Type	Private
Industry	Transport
Founded	1962
Headquarters	Gothenburg, Sweden
Key people	Gunnar Blomdahl (MD & COO)
Products	Ferries, port services, passenger transportation, freight transportation, holidays, business travel
Revenue	▲ 9,5 billion SEK (2006)
Employees	5,700 (2006)
Website	www.stenaline.com [1]

Stena Scandinavica in Kiel harbour

Stena Line is one of the world's largest ferry operators, with ferry services serving Scotland, Sweden, Northern Ireland, Denmark, Norway, England, Wales, Ireland, Germany, the Netherlands and Poland. Stena Line is a major unit of Stena AB, itself a part of the Stena Sphere, a grouping of Stena AB, Stena Metall AB and Stena Sessan AB. Stena Line also owns the Swedish vessels of the otherwise German- and Danish-owned Scandlines.

History

Stena Line was founded in, and is still operated from, Gothenburg, Sweden, by Sten A. Olsson when he acquired Skagenlinjen between Gothenburg, Sweden, and Fredrikshavn, Denmark, in 1962.

In 1972, Stena Line was one of the first ferry operators in Europe to introduce a computer-based reservation system for the travel business area. In 1978, the freight business area also started operating a computer-based reservation system.

The first freight-only route started between Gothenburg in Sweden and Kiel in Germany. The ship was the MS *Stena Transporter*.

During the 1980s, Stena acquired three other ferry companies. In 1981, Sessan Line, Stena's biggest competitor on Sweden–Denmark routes, was acquired and incorporated into Stena Line.[2] This included Sessan's two large newbuilds, *Kronprinsessan Victoria* and *Prinsessan Birgitta*, which became the largest ships operated by Stena by that date. In 1983, Stena acquired Varberg-Grenå Linjen, and two years later also the right to that company's former name, Lion Ferry. Lion Ferry continued as a separate marketing company until 1997, when it was incorporated into Stena Line.[3] In 1989, Stena acquired yet another ferry company, Stoomvaart Maatschappij Zeeland (SMZ) (which at the time traded under the name Crown Line). SMZ's Hoek van Holland–Harwich route then became a part of Stena Line.[4]

Stena Line doubled in size in 1990 with the acquisition of Sealink British Ferries from Sea Containers Ltd. This first became Sealink Stena Line, then Stena Sealink Line and finally Stena Line (UK), which now operates all of Stena's ferry services between Great Britain and Ireland.

Stena Jutlandica in Gothenburg

In 1996, Stena Line introduced its 20,000 tonne HSS (High-speed Sea Service) vessels, which operated from Belfast to Stranraer, Holyhead to Dún Laoghaire and Hoek van Holland to Harwich. In addition to the three 1500-passenger HSS vessels, Stena Line ordered two smaller 900-passenger HSS vessels to operate on the Gothenburg–Frederikshavn route. Due to the bankruptcy of the shipyard, only the first of these vessels was ever completed.

In 1998, Stena's operations from Dover and Newhaven were merged with P&O European Ferries to form P&O Stena Line, 40% of which was owned by Stena and 60% by P&O. In 2002, P&O acquired all of Stena's shares in the company, thus becoming the sole owner of P&O Stena Line, which soon changed its name to P&O Ferries.

In 2000, Stena Line purchased yet another Scandinavian ferry operator: Scandlines AB.

In November 2006, Stena ordered a pair of "super ferries" with a gross tonnage (GT) of 62,000 tons from Aker Yards Germany for delivery in 2010, with an option for two more ships of the same design.[5] The new ferries will be amongst the largest in the world,[6] and will be placed on Stena's North Sea route from Hoek van Holland to Harwich. The existing ships from the North Sea will be moved to the Kiel - Gothenborg route, whereas the ships from Kiel will move to the Gdynia to Karlskrona route. The new ferries are to be delivered in 2010, with the *Stena Hollandica* having entered service on May 16, 2010, and the *Stena Britannica* planned to enter service in the fall of 2010.[7] The company also moved its Belfast Terminal from Albert Quay to the new VT4 during May 2008. This has reduced the length of the crossing to Stranraer by 10 minutes.

In July 2009, Stena Line announced that it had repurchased its former ship, *Stena Parisien*, from SeaFrance. The ship is now known as the *Stena Navigator*. It had a comprehensive refit. Following on from this, the ship was introduced on the Stranraer to Belfast route, alongside HSS *Stena Voyager* and *Stena Caledonia*.

Irish Sea expansion

In December 2010, Stena Line announced it had acquired the Northern Irish operations of DFDS. The sale includes the Belfast to Heysham & Birkenhead routes, two vessels from the Heysham route (Scotia Seaways & Hibernia Seaways) and two chartered vessels from the Birkenhead route (*Mersey Seaways* & *Lagan Seaways*).[8] The Fleetwood to Larne route ended on 24 December 2010.

Routes

Irish Sea

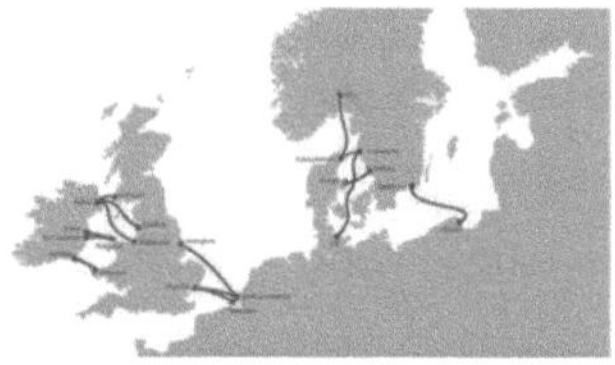
Stena Line route map

- Fishguard - Rosslare: *Stena Europe*
- Holyhead - Dublin Port: *Stena Adventurer*, *Stena Nordica*
- Holyhead - Dún Laoghaire: *Stena Explorer*
- Belfast - Cairnryan: *Stena Superfast VII*, *Stena Superfast VIII*
- Belfast - Heysham: *Stena Scotia*, *Stena Hibernia*
- Belfast - Birkenhead: *Stena Lagan*, *Stena Mersey*

North Sea

- Hoek van Holland - Harwich: *Stena Britannica*, *Stena Hollandica*
- Hoek van Holland - Killingholme: *Stena Transit*, *Stena Transporter*
- Rotterdam - Harwich: *Stena Freighter*, *Stena Carrier*

Scandinavia

- Göteborg - Fredrikshavn: *Stena Jutlandica*, *Stena Danica*, *Stena Carisma*, *Stena Scanrail*, *Stena Saga*, *Stena Carrier*
- Varberg - Grenaa: *Stena Nautica*
- Göteborg - Kiel: *Stena Germanica*, *Stena Scandinavica*, *Stena Freighter*
- Oslo - Fredrikshavn: *Stena Saga*
- Karlskrona - Gdynia: *Stena Spirit*, *Stena Vision*

Fleet

Current ships

Name	Built	Tonnage	Passengers	Notes
Stena Adventurer	2003	43,532	1,500	
Stena Baltica	1986	31,910	1,200	Sold to Jadrolinija
Stena Britannica	2010	63,039	1,200	
Stena Caledonia	1981	12,619	1,154	Laid up
Stena Carisma	1997	8,631	900	
Stena Carrier	1998	21,089	12	
Stena Danica	1983	15,899	2,300	
Stena Europe	1981	24,828	1,400	
Stena Explorer	1996	19,638	1,500	
Stena Feronia	1997	21,856	340	Currently out of service following collision with a cargo ship in Belfast Lough[9]
Stena Freighter	1998	21,004	12	
Stena Germanica	2001	44,237	900	
Stena Hibernia	1996	13,017	12	
Stena Hollandica	2010	63,039	1,200	
Stena Jutlandica	1996	29,691	1,500	
Stena Lagan	2005	26,500	980	
Stena Mersey	2005	26,500	980	
Stena Nautica	1986	11,763	663	
Stena Nordica	2000	24,206	405	
Stena Saga	1981	25,905	2,000	
Stena Scandinavica	2003	55,050	900	
Stena Scanrail	1973	7,504	12	
Stena Scotia	1996	13,600	12	
Stena Spirit	1988	42,448	2,200	
Stena Superfast VII	2001	30,285	1,200	Charted from Tallink. [10] "Stena charters Tallink vessels for Belfast run" [11]
Stena Superfast VIII	2001	30,285	1,200	Charted from Tallink.
Stena Transit	2011	33,690	300	
Stena Transporter	2011	33,690	300	
Stena Vision	1987	39,178	1,700	
Stena Voyager	1996	19,638	1,500	Laid up

Former ships

Further information: Stena Line's ship history

Gallery

Stena Traveller at Hoek van Holland

Stena Traveller at Hoek van Holland

References

[1] http://www.stenaline.com/
[2] "Sessan Linjen" (http://www.faktaomfartyg.nu/sessanlinjen.htm) (in Swedish). Fakta om Fartyg. . Retrieved 21 September 2007.
[3] "Lion Ferry" (http://www.faktaomfartyg.nu/lion_ferry.htm) (in Swedish). Fakta om Fartyg. . Retrieved 21 September 2007.
[4] "Stoomvaart Maatschappij Zeeland" (http://www.faktaomfartyg.nu/stoomvaart_maatschappij_zeeland.htm) (in Swedish). Fakta om Fartyg. . Retrieved 21 September 2007.
[5] "Press release 9. 11. 2006" (http://www.akeryards.com/?page=316&xml=A/134984/PR/200611/1087399). Aker Yards. . Retrieved 21 September 2007.
[6] "Stena Line announces major investment in innovative ferries" (http://www.stenaline.com/stena_line/stena_line_koncernen/media/en_gb/pressreleaser_18083.html). Stena Line press release. . Retrieved 21 September 2007.
[7] "'Superferry' in maiden Harwich crossing" (http://news.bbc.co.uk/2/hi/uk_news/england/essex/8685207.stm). BBC News. 16 May 2010. . Retrieved 16 May 2010.
[8] "DFDS sells Irish Sea ferry routes" (http://www.dfds.com/news/latest/20101202-1/). DFDS A/S. . Retrieved December 2010.
[9] The Guardian "Stena Line ferry collides with cargo ship in Belfast Lough" (http://www.guardian.co.uk/uk/2012/mar/08/stena-line-ferry-crash-belfast-lough), accessed 20 March 2012
[10] "Stena charters Tallink vessels for Belfast run" (http://marine-news.net/Gran_Tierra_Disappointed_by_Peru_Results_Updates_Colombia_Ops-i22491.html)
[11] http://marine-news.net/Gran_Tierra_Disappointed_by_Peru_Results_Updates_Colombia_Ops-i22491.html

External links

- Stena Line (http://www.stenaline.com/) - Corporate site
- Stena Line Reservations (http://www.stenaline.co.uk/) - UK online ticket reservations and home pages
- Stena Line Reservations (http://www.stenaline.se/) - Swedish online ticket reservations and home pages
- Stena Line Tourist Info (http://www.gothenburgguide.com/gothenburg-ferry-stena-line/) - General tourist information about Stena Line.

See also

- Stena Line Holland BV
- P&O Stena Line

'

MS *Superfast VII*, approaching Turku Repair Yard at Naantali, Finland.

Career	
Name:	***Stena Superfast VII***
Owner:	2002–2006: Superfast Ferries 2006 onwards: Baltic SF VII Ltd (Tallink)[1]
Operator:	2002–2006: Superfast Ferries 2006-2011: Tallink[1] 2011 onwards Stena Line[1]
Port of registry:	2002–2006: Piraeus, Greece 2006-2011: Paldiski, Estonia 2011 onwards: Belfast United Kingdom[1]
Route:	Belfast–Cairnryan
Builder:	HDW, Kiel, Germany
Yard number:	357[1]
Launched:	18 November 2000[1]
Acquired:	8 May 2001[1]
Maiden voyage:	17 May 2001[1]
In service:	17 May 2001[1]
Identification:	IMO number: 9198941[1]
Status:	In service
General characteristics (as built)[1]	
Class and type:	*Superfast VII* class fast ropax ferry
Tonnage:	30,285 GT

Displacement:	5,915 DWT
Length:	203.90 m (**unknown operator: u'strong'** ft)
Beam:	40.00 m (**unknown operator: u'strong'** ft)
Height:	40.0
Draught:	6.60 m ()
Decks:	10[2]
Ice class:	1 A Super[2]
Installed power:	4 × Wärtsilä-Sulzer NSD ZA V40S diesels combined 46000 kW
Propulsion:	2 propellers 3 bow thrusters 1 stern thruster[3]
Speed:	30.4 kn (**unknown operator: u'strong'** km/h)[2]
Capacity:	626 passengers 661 cars 1920 lane meters
Crew:	63[3]

MS ***Stena Superfast VII*** is a fast Ro-Pax ferry owned by the Estonian ferry company Tallink and chartered to Stena Line operated on their service between Belfast and Cairnryan. Built in 2001 by Howaldtswerke-Deutsche Werft (HDW) in Kiel, Germany for Attica Group's subsidiary Superfast Ferries, she was sold to her current owners in 2006.[1]

Concept and construction

The *Stena Superfast VII* was the first ship in a series of four ice-classified ropax ferries built by HDW in Kiel for Superfast Ferries' Baltic Sea services.[4] She was ordered in 1998, alongside sister ship MS *Superfast VIII*,[5] launched from dry dock on 8 November 2000 and was delivered to Superfast Ferries on 8 May 2001.[1]

Service history

2001–2006: Superfast Ferries

Following delivery the *Superfast VII* visited Rosyth, Scotland and Helsinki, Finland where she was displayed to the public. She inaugurated Superfast Ferries' Hanko (Finland) – Rostock (Germany) service on 17 May 2001. On 18 January 2002 a pregnant passenger went into labour on board while the *Superfast VII* was en route from Hanko to Rostock. Taking the ship to Karlshamn, Sweden where the expecting mother could be taken to a hospital was considered, but in the end the decision was made to bring a midwife on board by a pilot boat. The child had already been born by the time the midwife arrived on board, with the captain's wife assisting in childbirth.[1]

On 12 November 2004 the *Superfast VII* was entering Hanko harbour in heavy wind with the help of two tugs when she was grounded near the Hanko breakwater at 19:24 Eastern European Time. The grounding resulted in no major damage and the ship was able to continue to the harbour soon afterwards. 140 passengers were on board at the time of the incident. Subsequent enquiries revealed the Safety Management System instructions provided for the crew by Superfast Ferries had not included instructions for port steering during a storm, which had led to an incorrect estimation of the wind effect and insufficient utilization of the ship's navigational equipment.[3] Following the grounding the ship sailed to Luonnonmaan telakka in Naantali, Finland on 14 November 2004 and returned to service after repairs on 27 November 2004.[1]

2006-2011: Tallink

On 21 March 2006 Superfast Ferries sold their Baltic Sea operations (*Superfast VII*, *Superfast VIII* and *Superfast IX*) to the Estonia-based Tallink, with the delivery date set for 10 April 2006. According to the agreement Tallink could continue utilising the Superfast brand until the end of 2007 at latest.[1] [6] Following delivery to the new owners the ship was moved from Finnish to Estonian registry, and her route changed to Hanko–Paldiski–Rostock on 17 April 2006.[1] The route change caused problems however as Estonia was not a member of the Schengen Treaty (unlike Finland and Germany), and passport control facilities had to be built in all ports. Already in June of the same year the route reverted back to Hanko–Rostock. Around the same time the "Superfast" text on the ship's side was altered into "Superfast operated by Tallink".[1]

The route of the *Superfast VII* was changed to Helsinki–Rostock on 1 January 2007 and Tallinn–Helsinki–Rostock on 14 January 2007. Coinciding with this all remaining Superfast logos on the ships were painted over with Tallink logos, but otherwise the original Superfast livery was maintained.[1] Due to falling passenger numbers and rising fuel costs the route of *Superfast VII* and *VIII* will revert to Helsinki–Rostock in late 2008, while the *Superfast IX* will be chartered to Marine Atlantic in October 2008.[7] [8] Reportedly Tallink are considering the closure of the entire Tallink Superfast -division.[7] In January 2010, the *Superfast VII* and her sister were removed from service and laid up in Tallinn, but they resumed service on the Helsinki-Rostock route at the end of April 2010 and continuing during the summer and autumn season. On 29 and 30 December 2010, Superfast VII replaced MS Superstar on the route Tallinn-Helsinki due to scheduled docking[1]

2011 onwards: Stena Line

In March 2011, Stena Line announced they will be chartering the *Superfast VII* and sister ship *Superfast VIII*. The Superfast VII And Superfast VIII will launch the 21 November 2011 after major refit by MJM forces in Remontowa Shipyard in Poland[9]. The vessels will operate between Belfast and Stena Line's new terminal at Cairnryan.[10]

References

[1] Asklander, Micke. "M/S *Superfast VII* (2001)" (http://www.faktaomfartyg.nu/superfast_VII_2001.htm) (in Swedish). *Fakta om Fartyg*. . Retrieved 2008-04-02.

[2] "Tallink Superfast » Technical information" (http://www.tallinksilja.com/en/tab2/ships/superfast/technicalInformation/). Tallink official website. . Retrieved 2008-04-02.

[3] "MS *Superfast VII*, pohjakosketus Hangossa 12.11.2004" (http://www.onnettomuustutkinta.fi/uploads/qz40h.pdf) (in Finnish/Swedish/English) (PDF). Onnettomuustutkintakeskus. 2006-05-23. . Retrieved 2008-09-17.

[4] "Attica Receives New Superfast Ferry" (http://www.marinelink.com/Story/Attica+Receives+New+Superfast+Ferry-5557.html). *Marine Link.com*. Maritime Activity Reports, Inc.. 2001-05-09. . Retrieved 2008-09-17.

[5] "About Superfast Ferries" (http://www.superfastferry.com/). Superfast Ferries. . Retrieved 2008-09-17.

[6] "Sale of *Superfast VII, VIII* and *IX* to AS Tallink Grupp" (http://www.aferry.co.uk/news/viewpost.asp?news=94&Lang=UK). *Ferry News*. aferry.to. 2006-04-10. . Retrieved 2008-09-17.

[7] "Tallink harkitsee Superfast-pikalaivojen vetämistä Itämereltä" (http://www.hs.fi/talous/artikkeli/Tallink+harkitsee+Superfast-pikalaivojen+vetÃ¤mistÃ¤+ItÃ¤mereltÃ¤/1135237637365) (in Finnish). *HS.fi*. Helsingin Sanomat. 2008-07-23. . Retrieved 2008-09-17.

[8] "Charter of MS *Superfast IX*" (http://www.tallink.com/mainMenu/pressRoom/stockExchangeReleases/sr_25042008_sf9_charter.htm). *Tallink press release*. Tallink Grupp. 2008-04-25. . Retrieved 2008-09-17.

[9] "Two Stena Line chartered sister ferries come to Remontowa for major refit" (http://www.remontowa.com.pl/index.php?n=566). www.remontowa.com.pl. 2011. .

[10] "New ships for Scotland - NI service" (http://www.stenaline.co.uk/ferry/media/news/newships/). Stena Line. 7 March 2011. .

External links

- Superfast VII at marinetraffic.com (http://www.marinetraffic.com/ais/shipdetails.aspx?MMSI=276647000)

Article Sources and Contributors

' *Source*: http://en.wikipedia.org/w/index.php?title=MS_Stena_Superfast_VIII *Contributors*: AvicAWB, FJS15, FarbrorJoakim, Fyrael, Gene Nygaard, Haus, JonEastham, Kjet, Mbruce, Njm05, Philaweb, Philphos, PjotrMahh1, Roquai, Sladen, Traveletti, Tupsumato, Wwoods, Z10x, 6 anonymous edits

Tallinn *Source*: http://en.wikipedia.org/w/index.php?title=Tallinn *Contributors*: 1779Days, 195.80.96.xxx, 3 Löwi, A. B., Aaker, Abhijitsathe, Abu Amal Bahraini, Acarvin, Active Banana, Adam Carr, Ahoerstemeier, Ahtih, Al E., Al.Neuland, Alan.ca, Alastairgbrown, Alex Bakharev, AlexLL, Alexf, AlexiusHoratius, Alishah imtiaz, Amorfati00, Angr, Apokrif, Apuleuis damnius, Argo Navis, Arno, Aronian, Arpingstone, Arthur889, Attilios, Avala, Axt, Balcer, Baltic Seaman, BanRay, Baronnet, Bcnviajero, Beginning, Bhadani, Big Adamsky, Billinghurst, Biruitorul, BlackRaspberry, Blackorchidspider, Blue Elf, Bobo192, Bokpasa, Boothy443, Bpa, Brholden, Bronze Man of Tallinn, BuickCenturyDriver, BuzzWoof, CGP, CLW, CRKingston, Caponer, Cautious, Ceilican, Cemyildiz, Cessna1520, ChiLlBeserker, Chris97, Christian List, Clay, Clossius, Cmdrjameson, Colchicum, Colonies Chris, CommonsDelinker, Concledoc, Conversion script, Cray H, Curgny, Curps, D6, DMG413, DO'Neil, DVoit, Danielt998, Daviessimo, Dbenbenn, Deflective, Derek Ross, Deror avi, Digwuren, Docu, Doublex120, Dougweller, Dr. Blofeld, Dragon tomato, Drake600, Drunken Pirate, Durova, Ecuados, Editore99, Eeera, El C, Elekhh, Elockid, Epiphyllumlover, Eric82oslo, Ericdn, Espoo, EstGun1, Estlandia, Eugene van der Pijll, EugeneZelenko, EvanProdromou, ExRat, Ezhiki, Fang Aili, Feydey, Filzstift, Fireinacrowdedtheatre, Fisenko, Flrn, Flying Saucer, Folks at 137, Fonky, Fransvannes, Frisch1, Gadfium, Gareth Owen, Geniu, GeorgeLouis, Ghirlandajo, Gilgamesh, Gokhalevai, Graffoe, Gragox, Green Giant, Guliolopez, H2ppyme, Hanspostma, Harvestman, Haus, Herostratus, Holal5, Hu12, Hushpuckena, Hyperdoctor Phrogghrus, INkubusse, IainP, Ii345, Iifar, Isfisk, JIP, Jaan911, JaanMatti, Jaanusele, Jalexoid, Jamie C, Jared Preston, Jdforrester, JeR, Jemm, Jhendin, JinJian, Joey80, Johan Elisson, John K, Joldy, Jontts, Jose77, Joy, Jrcla2, Jreb01, K731, Kaganer, Kahkonen, Kaldenbach53, Kbdank71, Kennycard, Kibereke, Kneiphof, Knepflerle, Knutux, Koenigru, Korenyuk, Kostja, Kpalion, Kresspahl, Kristjan Parve, Kulmalukko, Kurivaim, Kwamikagami, LA2, Laur, Lear 21, Leonard G., Liftarn, LimoWreck, Lincher, Lipik, Lucyhill, Ludde23, M.K, M.nelson, MHatlie, MJCdetroit, Mack2, MacroDaemon, Magnus Manske, Maiko90, MaliNorway, Mammal4, Man vyi, Marathona, Mardus, Marek69, Mareklug, Mart wiki, Martpol, Matramas, Mbimmler, Mdann52, Metsavend, Meyousee, Mft87, Michael Gäbler, Michellecrisp, Mikehillman, Miki, Miller17CU94, Miraceti, Mister X, Mjroots, Mkjell, Mmwikiest, Moonraker, Mrtnspp, Munish.Phutela, Möchtegern, NATHANN901, NHRHS2010, NInTeNdO, Nach0king, Nakos2208, NawlinWiki, Nikai, Nima Baghaei, Nixer, Nk, Noblige, Noe, NotAnonymous, Nuffle, Nug, Numbo3, Nv8200p, Nyttend, Ohconfucius, Olessi, Olyus, Ott, Ottodv, OwenBlacker, PK2, Pablo-flores, Palmiro, PaulHanson, Pavel Vozenilek, Pb30, Pelmeen10, Perfectmiss, PeterThoeny, Petri Krohn, Phoenixdragon800, Picapica, Pigsonthewing, Plastikspork, Polylerus, PoorLeno, Poul818, Pt, Pudelek, PuzzletChung, Põhja Konn, Quibik, R9tgokunks, RJ CG, RSStockdale, Rain74, Random contributor, Raul1972, Reenem, Reinis, René Kedus, RexNL, Rey2, Rich Farmbrough, Riso, Rjwilmsi, Rocastelo, Rojomoke, Rotkraut, Rumpelstiltskin, Russavia, SI31, SJ Morg, SLWriter1120, Sander Säde, SchuminWeb, Seivad, Senzeichi, Ser Amantio di Nicolao, Serein (renamed because of SUL), Shoeofdeath, Shortfatlad, Sjc, Sky, Skäpperöd, Smith2006, Smjg, Snowdog, Sophievonanhalt, Sorent, Spacestoned, Splash, Sroen, Stan Shebs, Stepheng3, Suva, T-borg, Tallinn2011, Tamfang, Taneltammet, Tanuman, Tbhotch, Telempe, Teren, Termer, Texmon, The Anome, The wub, Thine Antique Pen, Thingg, Tibetan Prayer, Tide rolls, Tiigrikutzu, Tim!, Timsdad, Timvasquez, TintoRetto, Tkynerd, Tonusamuel, Torfkopp, Tpbradbury, Trojanavenger, Tuulispask, Tuupola, UltioUltionis, Ultratomio, Unigolyn, Valfontis, Varlaam, Vegaswikian, Verne Equinox, Vetalgo, Vic-vic, VicFromTheBlock, Victorian Chameleon, Victuallers, Vihelik, Vkem, Vongodric, Voyag, Vuo, Väino Koorberg, Völva, Wangi, Wecameasromans, Welsh, WhisperToMe, Wikiacc, Wikibofh, Wikid77, Wikipeditor, Wikiwolf ad, William Pietri, Wilson44691, WojPob, Wojtek C, XJamRastafire, YellowMonkey, Yuriybrisk, Zakuragi, Zotico, Zwanneman, £, Гатерас, 515 anonymous edits

Helsinki *Source*: http://en.wikipedia.org/w/index.php?title=Helsinki *Contributors*: -Anthony-, -Majestic-, -Marcus-, 159753, 28421u2232nfenfcenc, 334a, AB, Aaker, Abdullais4u, Abhijitsathe, Acdx, Achangeisasgoodasa, Acrajan, Addshore, Advaitrocking, Aecis, AgentCDE, Alansohn, Almost Anonymous, Alsandro, AndrewWTaylor, AntonioMartin, Antti Olavi Piirainen, Apalsola, Aphaia, ApprenticeFan, ArkansasTraveler, Atomichippo, Avala, Axeman89, Axilera, Axt, BKfi, BRUTE, Babylove10, Bbx, Beetstra, BenAveling, Bender235, Benoa, Birdie98, Bob247, Bpa, Breeezee, Brztm, C12H22O11, CGorman, CRKingston, CTajvar, Caltas, CambridgeBayWeather, Canderson7, Caponer, Captain Disdain, CaritaMarianne, Cburnett, Cccpike, Celithemis, Celticfan383, ChKa, Chavash, Chris the speller, Christian List, Chwech, Cimon Avaro, Clarifer, Clydog, Colinward1970, Colonies Chris, Colors99, Comidadehospital, Conscious, Conversion script, Corvus cornix, Courcelles, Cpettauer, Crazycraz525, Cwmhiraeth, DJ Clayworth, Daniel625, DanyWillis, Darin-0, Darius Dhlomo, Darklilac, Darrien, Darth Panda, Deb, Decltype, Defeatedfear, Delirium, Dendrer, DerHexer, DiiCinta, DocWatson42, Dolphin Jedi, Donarreiskoffer, Dr. Blofeld, DrKiernan, Dreg743, Drekscot, Drfrisby, Drmies, Durin, Dwstein, DéRahier, Edmund1989, Edward, El C, Elekhh, Emil Kastberg, Enchanter, Erianna, Erik Kennedy, Espoo, Euchiasmus, Eugene van der Pijll, Evanreyes, Excirial, Fab, Fanghong, Faradayplank, FernEn, Fieldday-sunday, Flami72, Flockofpidgeons, Flrn, FlyerBoy, Fourdee, Free11, Frøde Næss, Gaff, Garik 11, George Esayas, Gertjan R., GilbertoSilvaFan, Giler, Gilgamesh, Gillis, Gogo Dodo, Graaf, Graham87, Green Giant, Greenshed, Grutness, Gsklee, Guaka, Guidod, Guillom, Gxojo, HTR233, HamYoyo, Hapsiainen, Harjasusi, Hartz, Hayden120, Hellbus, Heroblue, Hippophaë, Hirvio, Hmains, Humus sapiens, Hurmata, Hydrox, IanManka, Ilmari Karonen, Im a Socialist! What Are You, Immaman, Imperium Europeum, InvictaHOG, Irate velociraptor, Ivarercool, Ixfd64, J.delanoy, JAM67, JGHowes, JIP, JaanMatti, Jarri K, JdeJ, JeLuF, JeffyJeffyMan2004, Jeppiz, JeremyA, Jeronimo, Jhendin, Jimp, JinJian, Jkstark, Joey80, Jogers, JonMoore, Jonik, JonneeS, Jontts, Jose77, Jpatokal, JtTractorBoy, Juhanix, Junttijuntus-, Jyril, KFP, Kahvipannu, Kaiba, KapilTagore, Karmosin, Katalla, Kateshortforbob, Keilana, Kevin B12, Khoikhoi, KihoDeal, Kinkku Ananas, Kitia, Kjet, Knutux, Koenigru, Kpjas, Kwamikagami, LA RoeDoe, Lacaid, Lalli, Lapinmies, Lastflyingman, Laura Elizabeth, Laurens-af, Lear 21, Leopea, Levil, LibStar, Liftarn, Lightmouse, Lights, Litegrann, Logan, Lom Konkreta, Longeness, Looxix, Lotje, Lucky Juice, Lucyhill, Lupin, M.nelson, MFantti, MJCdetroit, Mahamoty, Malcolm, Malickfan86, Mani1, Manuel Anastácio, Marcika, Marco polo, Marcus2, Marek69, Markonen, Markpeak, Mashutik, Massagetae, Matissek, Matthew Brandon Yeager, Maya, Mayhaymate, Mboverload, Mcbouncy, Mcphersonm80, MegX, Mentifisto, Mezaco, Mherlo, Mic, Middayexpress, Midgrid, Migro, Mikko, Mikko-Petteri, Miksa76, Millisits, Mipeltaja, Mirkonicholson, Mitte, MllandryQC, Mnh123, MoRsE, Moisesencyclopedia, Monegasque, Morganshade, Mouse is back, Muad, Multixfer, MusicGeek101, Mwanner, Mynthon, N96, NInTeNdO, NMS, NawlinWiki, Neilbeach, Neurovelho, Nick, NigelR, Nikai, Nirro, Nivix, Nk, Nn123645, Node ue, Nortonyodles, Numbo3, Oghmoir, Olborne, Olyus, Orange Suede Sofa, OrangeDog, Oreo Priest, Otriis, Oxymoron83, PBP, Pandion auk, Patrick, PatrikR, Paulinho28, Pavel Vozenilek, Pe3, PerLundberg, Perfectmiss, Pertsaboy, Pethr, Petri Krohn, Phantomsteve, Philly boy92, Philthecow, Pinkadelica, Pitcairn66, Pmcm, Popeda3, Pudeo, Purple, Purpleturple, Q, Querto, Quuxplusone, RG2, RainbowOfLight, Rallette, Red Winged Duck, Rendeath, Renne, RevRagnarok, RexNL, Rhollenton, Rich Farmbrough, Richard Harvey, Richmeister, Rjd0060, Rjwilmsi, RobertG, Robfwoods, Roger Roger, Romanm, RonDivine, Ruhrjung, Rune.welsh, SDC, SF94131, SI31, SageVita, Sakuliini, Sam Hocevar, Samtheboy, Saruwine, Sasajid, SchfiftyThree, Schmiteye, Scotthatton, Screamin2u, Sebjarod, Secretlondon, Sertmann, Seth Ilys, Shadowjams, Shanes, Shawn in Montreal, Shikker, Siipikarja, Silvonen, SimonP, Sjc07, Sjorford, Skaj, Skeeter08865, Skinnyweed, Skrim, Smilinghobbes, Snellman, Snowgrouse, Snowolf, Soliloquial, SpookyMulder, SqueakBox, Ssilvers, Ssipila, Stadscykel, Stemonitis, Stephen, Stephenchou0722, SteveDay, Stig Blomqvist, Subdependent, Superfluid, Sveeger, Synapsi, TPjalk, Tabletop, TarmoK, Tarnas, Tbackstr, Tbhotch, Teemu Leisti, Template namespace initialisation script, ThatRockMetalGuy, The Rambling Man, The Universe Is Cool, The Wild West guy, The wub, The0ther, Thermos, Thuresson, Tide rolls, Tim!, Timir2, Timo Honkasalo, Tnxman307, Tobby72, Tonym88, Totti76, Tpbradbury, Tpienonen, Trainthh, Tsemii, Tsujigiri, Tttammin, Tuomas, Turkuun, Tuuraan78, UFOPOLI, Ultratomio, Urhixidur, User F203, Username20111123, VKing, Viajero, Victor Chmara, Ville V. Kokko, Virtakuono, Vkem, Vovan7349, Vuo, Völva, Wadems, Wai Hong, Walkerma, Wavelength, Wayne Miller, Welsh, WhisperToMe, Wiglaf, Wikipeditor, Wimt, Winchelsea, WojPob, Writtenright, Xionbox, Yupik, Zakuragi, Zam16, Zbcc, Zedkoman, Zocky, Zoicon5, Zserango, ஹெலெசின்கி லொனொரந்து, 985 anonymous edits

Rostock *Source*: http://en.wikipedia.org/w/index.php?title=Rostock *Contributors*: *drew, 52 Pickup, Adam sk, Afrlmmr, Agathoclea, Ahoerstemeier, Anchjo, Andrwsc, Angr, Anthonyrussell84, Athrash, Attilios, Axeman89, Axt, BD2412, Balcer, Baldhur, Barliner, Baxter9, Bellcristi, Bentley4, Bermicourt, Bernhardwiki, Bevo, Bissinger, C2312, CRKingston, Cautious, Charles01, Chep87, Civvi, Cmdrjameson, Colonel Mustard, Cyhatch, D6, DO'Neil, Daderot, Darkone, DerBorg, Dewritech, Dheerav2, Diannaa, Ehrlich91, Elmju, Erik Kennedy, FBrei, Frokor, G716, Gaius Cornelius, Good Olfactory, Grand-Duc, Grandiose, GreatWhiteNortherner, Heah, HerkusMonte, Hixteilchen, Hmains, Hu12, Hughstew, Hvn0413, Icairns, Ilian tr, IntheWork, JBellis, JDavid, JPopovic, Jim Michael, Jimbo Herndan, John, Kaarinahollo, Kasp2008, Kejo13, Kingjeff, Knepflerle, Krawi, Kresspahl, Krzysztoflew, Kuususi, Kwamikagami, Kyle1278, LIU, LThF, LeVoyageur, Lear 21, Leifern, Leolaursen, Lightmouse, LuWe, Mais oui!, Man vyi, Marek69, Markussep, Martarius, MatthiasHuehr, Mervyn, Mm pie, Moixa, Monorgan, MrsMyer, Net-net, Nskillen, Nycgreg, Olivier, Owen, OwenBlacker, Oxydo, PGWG, PdFsoap, PeterHuntington, Pischdi Hufnagel, Policron, Puffin, R'n'B, RSStockdale, RedWolf, RexNL, Rimshot, Rjwilmsi, Rlquall, Royalmate1, S.K., Sejtam, Sky, Skäpperöd, Sleigh, Ssudo2002, Stemonitis, SteveMcP, Stullkowski, Tancred, Timwi, Tpbradbury, Tufao, Varlaam, Vedran12, Vslashg, Wee Jimmy, WhiteCrow, Whitejay251, Yolgustada, 227 anonymous edits

' *Source*: http://en.wikipedia.org/w/index.php?title=MS_Star *Contributors*: Ahsoous, Anttihav, Apalsola, AvicAWB, FarbrorJoakim, Haus, JonEastham, Kjet, MBK004, Mona, Philaweb, PjotrMahh1, Rich Farmbrough, Sladen, Tupsumato, Wwoods, 4 anonymous edits

Paldiski *Source*: http://en.wikipedia.org/w/index.php?title=Paldiski *Contributors*: Amherst99, Arkid77, Beeblebrox5000, Blue Elf, Cleverboy, Cray H, Dr. Blofeld, Flying Saucer, Jeck2009, Jimjamjak, Jrc, K731, Kallemax, Kulmalukko, Lunarmys, MJCdetroit, Marek69, Mcgon1979, Mitrius, Mmh, Mr.Z-man, Orderinchaos 2, Philaweb, Picapica, Reimgild, Rune X2, Sardanaphalus, Senzeichi, Tec15, Timothy Titus, Van der Hoorn, Whhalbert, Xiao yang, 20 anonymous edits

Stena_Line *Source*: http://en.wikipedia.org/w/index.php?title=Stena_Line *Contributors*: Aaronjhill, Ab irato, Abberley2, Analoguedragon, Andemalm, AndersL, Andrewahere, AvicAWB, Bardsandwarriors, Bbx, Biruitorul, BrownHairedGirl, C628, CHacker, Chochopk, Choster, Chris0693, Classical geographer, Cmdrjameson, DOSGuy, Darklilac, Dd1129, Dexxa90cc, Dom0803, DrFrench, Dsanders12, Eatcacti, Fabiform, Fbeauregard, Finavon, FiringRange, GTlikeitmatters, Geo Swan, Goswrang, H1523702, Hailey C. Shannon, Hammersfan, Haus, IRI5HJ4CK, JBellis, Jamie2k9, Janx Spirit, JoanneB, JohJak2, Kalaha, Keys767, Kjet, Kristof vt, Ksll, Lasuisflavus, Leinad, LiverpoolFitzy2009, MRSC, MacSteaphain, Mais oui!, Mark83, Mbruce, Mild Bill Hiccup, Mion, Mrceleb2007, Nach0king, Natox, Niall123, Njm05, NotMuchToSay, Nurg, OmegaRED, Paperones, Parsecboy, Philphos, Plasma east, Prackman, Quistnix, Rapido, Red Fox, Roquai, RyanM96, SFC9394, Sealynx, Siswrn, Skier Dude, Skinsmoke, Sladen, Tassedethe, Trevor Marron, Trident13, Vegaswikian, Velella, Wayneamlwch, Welkinridge, Wexcan, WhisperToMe, Yeti9389, 145 anonymous edits

' *Source*: http://en.wikipedia.org/w/index.php?title=MS_Stena_Superfast_VII *Contributors*: AvicAWB, Chris the speller, FJS15, FarbrorJoakim, Fyrael, Gourleyo, Haus, JonEastham, Joseph S 97, Kjet, Mbruce, Njm05, Philaweb, Philphos, PjotrMahh1, Re465, Roquai, Saberwyn, Sailing to Byzantium, Setti21, Sladen, Traveletti, Tsange, Tupsumato, Wwoods, 17 anonymous edits

Image Sources, Licenses and Contributors

Image:MS Superfast VIII.jpg *Source*: http://en.wikipedia.org/w/index.php?title=File:MS_Superfast_VIII.jpg *License*: unknown *Contributors*: User:Rovp

File:Flag of Greece.svg *Source*: http://en.wikipedia.org/w/index.php?title=File:Flag_of_Greece.svg *License*: unknown *Contributors*: (of code) (talk)

File:Flag of Estonia.svg *Source*: http://en.wikipedia.org/w/index.php?title=File:Flag_of_Estonia.svg *License*: unknown *Contributors*: User:PeepP, User:SKopp

File:Civil Ensign of the United Kingdom.svg *Source*: http://en.wikipedia.org/w/index.php?title=File:Civil_Ensign_of_the_United_Kingdom.svg *License*: unknown *Contributors*: User:Pumbaa80

File:Tallinncollage3.jpg *Source*: http://en.wikipedia.org/w/index.php?title=File:Tallinncollage3.jpg *License*: unknown *Contributors*: User:Mmwikiest

File:Flag of Tallinn.svg *Source*: http://en.wikipedia.org/w/index.php?title=File:Flag_of_Tallinn.svg *License*: unknown *Contributors*: User:Wikiacc

File:Tallinn greater coatofarms.png *Source*: http://en.wikipedia.org/w/index.php?title=File:Tallinn_greater_coatofarms.png *License*: unknown *Contributors*: Bryan Derksen, Bvs-aca, Darwinius, Joey-das-WBF, LA2, Nouill, Rain74, Senzeichi, Valentinian, WikedKentaur

file:Estonia location map.svg *Source*: http://en.wikipedia.org/w/index.php?title=File:Estonia_location_map.svg *License*: unknown *Contributors*: User:NordNordWest

File:Red pog.svg *Source*: http://en.wikipedia.org/w/index.php?title=File:Red_pog.svg *License*: unknown *Contributors*: Anomie

File:Flag of et-Harju maakond.svg *Source*: http://en.wikipedia.org/w/index.php?title=File:Flag_of_et-Harju_maakond.svg *License*: unknown *Contributors*: Homo lupus, Urmas, Valentinian, Walden69, WikedKentaur, 1 anonymous edits

File:Coat of arms of Tallinn (small).svg *Source*: http://en.wikipedia.org/w/index.php?title=File:Coat_of_arms_of_Tallinn_(small).svg *License*: unknown *Contributors*: BrightRaven, Joey-das-WBF, Madden, Nouill, Urmas, Valentinian

File:Reval-8371.jpg *Source*: http://en.wikipedia.org/w/index.php?title=File:Reval-8371.jpg *License*: unknown *Contributors*: User:Hedwig Storch

File:Danmarks flag 1219 Lorentzen.jpg *Source*: http://en.wikipedia.org/w/index.php?title=File:Danmarks_flag_1219_Lorentzen.jpg *License*: unknown *Contributors*: user:Kooma

File:Tallinn-panorama.jpg *Source*: http://en.wikipedia.org/w/index.php?title=File:Tallinn-panorama.jpg *License*: unknown *Contributors*: Ralf Roletschek (user Ralf Roletschek at de.wikipedia) Later versions were uploaded by Smial at de.wikipedia.

File:Revals segl.svg *Source*: http://en.wikipedia.org/w/index.php?title=File:Revals_segl.svg *License*: unknown *Contributors*: Illustration by Professor Julius Magnus Petersen (4 Sep 1827 – 1 Feb 1917). SVG created by Termer

File:Alexey Bogolybov - Port of Tallinn (1853).jpg *Source*: http://en.wikipedia.org/w/index.php?title=File:Alexey_Bogolybov_-_Port_of_Tallinn_(1853).jpg *License*: unknown *Contributors*: Ahsoous

File:VanaToomas1530.jpg *Source*: http://en.wikipedia.org/w/index.php?title=File:VanaToomas1530.jpg *License*: unknown *Contributors*: User:Terker

File:Raekoja plats (Tallinn).jpg *Source*: http://en.wikipedia.org/w/index.php?title=File:Raekoja_plats_(Tallinn).jpg *License*: unknown *Contributors*: Holger Vaga from Tallinn, Estonia

File:Magnify-clip.png *Source*: http://en.wikipedia.org/w/index.php?title=File:Magnify-clip.png *License*: unknown *Contributors*: User:Erasoft24

File:TallinnPan.jpg *Source*: http://en.wikipedia.org/w/index.php?title=File:TallinnPan.jpg *License*: unknown *Contributors*: User:Terker

File:Districts of Tallinn.jpg *Source*: http://en.wikipedia.org/w/index.php?title=File:Districts_of_Tallinn.jpg *License*: unknown *Contributors*: User:Hannu

File:Tallinn-Tornimae.jpg *Source*: http://en.wikipedia.org/w/index.php?title=File:Tallinn-Tornimae.jpg *License*: unknown *Contributors*: Flying Saucer, JaanMatti, Quibik, Shell Kinney

File:Rotermanni.IMG_2809.JPG *Source*: http://en.wikipedia.org/w/index.php?title=File:Rotermanni.IMG_2809.JPG *License*: unknown *Contributors*: User:Avjoska

File:Tallinna_inglise_kolledži_spordihoone.JPG *Source*: http://en.wikipedia.org/w/index.php?title=File:Tallinna_inglise_kolledži_spordihoone.JPG *License*: unknown *Contributors*: User:Avjoska

File:Reval-8328.jpg *Source*: http://en.wikipedia.org/w/index.php?title=File:Reval-8328.jpg *License*: unknown *Contributors*: User:Hedwig Storch

File:Tallinn-2007-rr-125.jpg *Source*: http://en.wikipedia.org/w/index.php?title=File:Tallinn-2007-rr-125.jpg *License*: unknown *Contributors*: User:Ralf Roletschek

File:St Olaf's church, Tallinn, July 2008.jpg *Source*: http://en.wikipedia.org/w/index.php?title=File:St_Olaf's_church,_Tallinn,_July_2008.jpg *License*: unknown *Contributors*: Olga Itenberg

File:StadtmauerTallinn.jpg *Source*: http://en.wikipedia.org/w/index.php?title=File:StadtmauerTallinn.jpg *License*: unknown *Contributors*: Hajotthu 18:19, 29. Apr. 2008 (CEST). Original uploader was Hajotthu at de.wikipedia

File:Kadrioru loss,.jpg *Source*: http://en.wikipedia.org/w/index.php?title=File:Kadrioru_loss,.jpg *License*: unknown *Contributors*: User:Ren12

File:Viru Gates, IMG 4605.jpg *Source*: http://en.wikipedia.org/w/index.php?title=File:Viru_Gates,_IMG_4605.jpg *License*: unknown *Contributors*: Karen P

File:StCatherinesPassage.jpg *Source*: http://en.wikipedia.org/w/index.php?title=File:StCatherinesPassage.jpg *License*: unknown *Contributors*: Flying Saucer

File:Orthodox Alexander Nevsky Cathedral in Tallinn.jpg *Source*: http://en.wikipedia.org/w/index.php?title=File:Orthodox_Alexander_Nevsky_Cathedral_in_Tallinn.jpg *License*: unknown *Contributors*: User:Julo

File:Tallinn-2007-rr-106.jpg *Source*: http://en.wikipedia.org/w/index.php?title=File:Tallinn-2007-rr-106.jpg *License*: unknown *Contributors*: User:Ralf Roletschek

File:Tallinn-2007-rr-108.jpg *Source*: http://en.wikipedia.org/w/index.php?title=File:Tallinn-2007-rr-108.jpg *License*: unknown *Contributors*: User:Ralf Roletschek

File:Tallinn Schwarzhaupterhaus.jpg *Source*: http://en.wikipedia.org/w/index.php?title=File:Tallinn_Schwarzhaupterhaus.jpg *License*: unknown *Contributors*: Original uploader was Brunswyk at de.wikipedia (Original text :)

File:Pirita kloostri varemed kalmistuga.jpg *Source*: http://en.wikipedia.org/w/index.php?title=File:Pirita_kloostri_varemed_kalmistuga.jpg *License*: unknown *Contributors*: User:Ren12

File:Estonia 1456 - Sing to defeat a enemy...jpg *Source*: http://en.wikipedia.org/w/index.php?title=File:Estonia_1456_-_Sing_to_defeat_a_enemy...jpg *License*: unknown *Contributors*: Dennis Jarvis

File:Tallinns Airport.jpg *Source*: http://en.wikipedia.org/w/index.php?title=File:Tallinns_Airport.jpg *License*: unknown *Contributors*: Karabinier (talk) 14:51, 12 June 2008 (UTC)

File:Copterline in Tallinn, HolgerVaga.jpg *Source*: http://en.wikipedia.org/w/index.php?title=File:Copterline_in_Tallinn,_HolgerVaga.jpg *License*: unknown *Contributors*: Holger Vaga from Tallinn, Estonia

File:Flag of Finland.svg *Source*: http://en.wikipedia.org/w/index.php?title=File:Flag_of_Finland.svg *License*: unknown *Contributors*: User:SKopp

File:Flag of Åland.svg *Source*: http://en.wikipedia.org/w/index.php?title=File:Flag_of_Åland.svg *License*: unknown *Contributors*: User:Túrelio

File:Flag of Sweden.svg *Source*: http://en.wikipedia.org/w/index.php?title=File:Flag_of_Sweden.svg *License*: unknown *Contributors*: Anomie

File:Flag of Germany.svg *Source*: http://en.wikipedia.org/w/index.php?title=File:Flag_of_Germany.svg *License*: unknown *Contributors*: Anomie

File:Flag of Russia.svg *Source*: http://en.wikipedia.org/w/index.php?title=File:Flag_of_Russia.svg *License*: unknown *Contributors*: Anomie

File:Morning_in_Tallinn_Harbor.jpg *Source*: http://en.wikipedia.org/w/index.php?title=File:Morning_in_Tallinn_Harbor.jpg *License*: unknown *Contributors*: Ilya

File:Baltic Queen Tallinn 2009-04-23.JPG *Source*: http://en.wikipedia.org/w/index.php?title=File:Baltic_Queen_Tallinn_2009-04-23.JPG *License*: unknown *Contributors*: User:Kjet

File:Tallinnharbour09.png *Source*: http://en.wikipedia.org/w/index.php?title=File:Tallinnharbour09.png *License*: unknown *Contributors*: User:René Kedus

File:Atlantica vs Classica.jpg *Source*: http://en.wikipedia.org/w/index.php?title=File:Atlantica_vs_Classica.jpg *License*: unknown *Contributors*: n3zwork

File:Tallin train station.JPG *Source*: http://en.wikipedia.org/w/index.php?title=File:Tallin_train_station.JPG *License*: unknown *Contributors*: Cmelak770, Flying Saucer, G.dallorto, Palefire, WikedKentaur, 1 anonymous edits

File:Flag of the United States.svg *Source*: http://en.wikipedia.org/w/index.php?title=File:Flag_of_the_United_States.svg *License*: unknown *Contributors*: Anomie

File:Flag of Poland.svg *Source*: http://en.wikipedia.org/w/index.php?title=File:Flag_of_Poland.svg *License*: unknown *Contributors*: Anomie, Mifter

File:Flag of the United Kingdom.svg *Source*: http://en.wikipedia.org/w/index.php?title=File:Flag_of_the_United_Kingdom.svg *License*: unknown *Contributors*: Anomie, Good Olfactory, Mifter

File:Flag of Belgium (civil).svg *Source*: http://en.wikipedia.org/w/index.php?title=File:Flag_of_Belgium_(civil).svg *License*: unknown *Contributors*: Bean49, David Descamps, Dbenbenn, Denelson83, Evanc0912, Fry1989, Gabriel trzy, Howcome, IvanOS, Ms2ger, Nightstallion, Oreo Priest, Rocket000, Rodejong, Sir Iain, ThomasPusch, Warddr, Zscout370, 4 anonymous edits

File:Flag of the Netherlands.svg *Source*: http://en.wikipedia.org/w/index.php?title=File:Flag_of_the_Netherlands.svg *License*: unknown *Contributors*: User:Zscout370

File:Flag of Latvia.svg *Source*: http://en.wikipedia.org/w/index.php?title=File:Flag_of_Latvia.svg *License*: unknown *Contributors*: User:SKopp

File:Flag of Italy.svg *Source*: http://en.wikipedia.org/w/index.php?title=File:Flag_of_Italy.svg *License*: unknown *Contributors*: Anomie

File:Flag of Lithuania.svg *Source*: http://en.wikipedia.org/w/index.php?title=File:Flag_of_Lithuania.svg *License*: unknown *Contributors*: User:SKopp
File:HelsinkiMontage NoEffects.jpg *Source*: http://en.wikipedia.org/w/index.php?title=File:HelsinkiMontage_NoEffects.jpg *License*: unknown *Contributors*: User:Acdx
File:Helsinki.vaakuna.svg *Source*: http://en.wikipedia.org/w/index.php?title=File:Helsinki.vaakuna.svg *License*: unknown *Contributors*: User:-Majestic-
file:Finland location map.svg *Source*: http://en.wikipedia.org/w/index.php?title=File:Finland_location_map.svg *License*: unknown *Contributors*: User:NordNordWest
File:Helsinki 1820.jpg *Source*: http://en.wikipedia.org/w/index.php?title=File:Helsinki_1820.jpg *License*: unknown *Contributors*: Drawn by Carl Ludvig Engel
File:Helsinki SPOT 1021.jpg *Source*: http://en.wikipedia.org/w/index.php?title=File:Helsinki_SPOT_1021.jpg *License*: unknown *Contributors*: User:Spot Image
File:Suomenlinna.jpg *Source*: http://en.wikipedia.org/w/index.php?title=File:Suomenlinna.jpg *License*: unknown *Contributors*: User:Michal.pise
File:Aleksanterinkatu Helsinki summer.jpg *Source*: http://en.wikipedia.org/w/index.php?title=File:Aleksanterinkatu_Helsinki_summer.jpg *License*: unknown *Contributors*: User:Mahlum
File:Aleksi Helsinki Market Street.jpg *Source*: http://en.wikipedia.org/w/index.php?title=File:Aleksi_Helsinki_Market_Street.jpg *License*: unknown *Contributors*: User:Axilera
File:HelsinkiToolonlahti.jpg *Source*: http://en.wikipedia.org/w/index.php?title=File:HelsinkiToolonlahti.jpg *License*: unknown *Contributors*: Author: Matthew McPherson Helsinki, Finland
File:Helsinki Lutheran Chathedral and the statue.jpg *Source*: http://en.wikipedia.org/w/index.php?title=File:Helsinki_Lutheran_Chathedral_and_the_statue.jpg *License*: unknown *Contributors*: User:-Majestic-
File:ViewOfficeParliamentFinland.jpg *Source*: http://en.wikipedia.org/w/index.php?title=File:ViewOfficeParliamentFinland.jpg *License*: unknown *Contributors*: User:Thermos
File:Cafe Recatta.jpg *Source*: http://en.wikipedia.org/w/index.php?title=File:Cafe_Recatta.jpg *License*: unknown *Contributors*: User:Axilera
File:Southern Helsinki panorama 2011-06-28 1.jpg *Source*: http://en.wikipedia.org/w/index.php?title=File:Southern_Helsinki_panorama_2011-06-28_1.jpg *License*: unknown *Contributors*: User:KFP
File:Helsinki_City_Center.jpg *Source*: http://en.wikipedia.org/w/index.php?title=File:Helsinki_City_Center.jpg *License*: unknown *Contributors*: User:Axilera
File:Kamppi_Center_Panoramic.jpg *Source*: http://en.wikipedia.org/w/index.php?title=File:Kamppi_Center_Panoramic.jpg *License*: unknown *Contributors*: User:Axilera
File:Kamppi Center II.jpg *Source*: http://en.wikipedia.org/w/index.php?title=File:Kamppi_Center_II.jpg *License*: unknown *Contributors*: User:Mahlum
File:Helsingin yliopiston päärakennus.jpg *Source*: http://en.wikipedia.org/w/index.php?title=File:Helsingin_yliopiston_päärakennus.jpg *License*: unknown *Contributors*: Apalsola, Fransvannes, Hautala, Tomisti
File:Helia 2005-10-05.jpg *Source*: http://en.wikipedia.org/w/index.php?title=File:Helia_2005-10-05.jpg *License*: unknown *Contributors*: Jannex
File:Helsinki Kiasma.jpg *Source*: http://en.wikipedia.org/w/index.php?title=File:Helsinki_Kiasma.jpg *License*: unknown *Contributors*: User:ZeroOne
File:2005 World Championships in Athletics 2.jpg.JPG *Source*: http://en.wikipedia.org/w/index.php?title=File:2005_World_Championships_in_Athletics_2.jpg.JPG *License*: unknown *Contributors*: Apalsola, Arch2all, Lumijaguaari, Pyhajumbo, Tomisti
File:Helsinki region roads.jpg *Source*: http://en.wikipedia.org/w/index.php?title=File:Helsinki_region_roads.jpg *License*: unknown *Contributors*: User:Migro
File:Helsinki Metro train.jpg *Source*: http://en.wikipedia.org/w/index.php?title=File:Helsinki_Metro_train.jpg *License*: unknown *Contributors*: Lussmu
File:Helsinki-Malmi 2008-001.jpg *Source*: http://en.wikipedia.org/w/index.php?title=File:Helsinki-Malmi_2008-001.jpg *License*: unknown *Contributors*: User:BKfi
File:Skyline of Helsinki as seen from the Erottaja fire station.jpg *Source*: http://en.wikipedia.org/w/index.php?title=File:Skyline_of_Helsinki_as_seen_from_the_Erottaja_fire_station.jpg *License*: unknown *Contributors*: User:JIP
File:Helsinki Railway Station 20050604.jpg *Source*: http://en.wikipedia.org/w/index.php?title=File:Helsinki_Railway_Station_20050604.jpg *License*: unknown *Contributors*: User:Revontuli
File:Helsinkimerelta--GFDL--.jpg *Source*: http://en.wikipedia.org/w/index.php?title=File:Helsinkimerelta--GFDL--.jpg *License*: unknown *Contributors*: Tpienonen at en.wikipedia
File:FinlandNationalMuseum.jpg *Source*: http://en.wikipedia.org/w/index.php?title=File:FinlandNationalMuseum.jpg *License*: unknown *Contributors*: Fransvannes, Mahlum, Samulili, Thermos
File:Senaatintori joulukuisena aamuna 2004.jpg *Source*: http://en.wikipedia.org/w/index.php?title=File:Senaatintori_joulukuisena_aamuna_2004.jpg *License*: unknown *Contributors*: User:Jonik
File:Helsinki Pohjoisranta.jpg *Source*: http://en.wikipedia.org/w/index.php?title=File:Helsinki_Pohjoisranta.jpg *License*: unknown *Contributors*: FlickreviewR, Jensens, Nenne, Petri Krohn
File:Stockmann department store in Helsinki at dawn in December 2004.jpg *Source*: http://en.wikipedia.org/w/index.php?title=File:Stockmann_department_store_in_Helsinki_at_dawn_in_December_2004.jpg *License*: unknown *Contributors*: User:Jonik
File:Mittsommernacht-in-helsinki.jpg *Source*: http://en.wikipedia.org/w/index.php?title=File:Mittsommernacht-in-helsinki.jpg *License*: unknown *Contributors*: Marcela (Ralf Roletschek)
File:Helsinki_Market.JPG *Source*: http://en.wikipedia.org/w/index.php?title=File:Helsinki_Market.JPG *License*: unknown *Contributors*: Apalsola, Fransvannes, Mitte, Väsk
File:Helsinki center evening.jpg *Source*: http://en.wikipedia.org/w/index.php?title=File:Helsinki_center_evening.jpg *License*: unknown *Contributors*: User:Sergey Tchernyakov
File:Tower of the Helsinki Olympic Stadium.jpg *Source*: http://en.wikipedia.org/w/index.php?title=File:Tower_of_the_Helsinki_Olympic_Stadium.jpg *License*: unknown *Contributors*: User:Jonik
File:Hesakirkot--GFDL--.jpg *Source*: http://en.wikipedia.org/w/index.php?title=File:Hesakirkot--GFDL--.jpg *License*: unknown *Contributors*: Shyam, Tpienonen
File:Uspenski_Cathedral_Helsinki.jpg *Source*: http://en.wikipedia.org/w/index.php?title=File:Uspenski_Cathedral_Helsinki.jpg *License*: unknown *Contributors*: user:InvictaHOG
File:FinlandParliament.jpg *Source*: http://en.wikipedia.org/w/index.php?title=File:FinlandParliament.jpg *License*: unknown *Contributors*: User:Thermos
File:Esplanadin puisto.jpg *Source*: http://en.wikipedia.org/w/index.php?title=File:Esplanadin_puisto.jpg *License*: unknown *Contributors*: Gillis, Mahlum, Nironen, 1 anonymous edits
File:Finnish National Theatre.jpg *Source*: http://en.wikipedia.org/w/index.php?title=File:Finnish_National_Theatre.jpg *License*: unknown *Contributors*: Apalsola, Courcelles, Estormiz, EugeneZelenko, Fransvannes, Lumijaguaari, Mahlum, Skorpion87, Tomisti
File:Temppeliaukio Church 3.jpg *Source*: http://en.wikipedia.org/w/index.php?title=File:Temppeliaukio_Church_3.jpg *License*: unknown *Contributors*: Apalsola, Pertsaboy
File:Hietaniemi beach.jpg *Source*: http://en.wikipedia.org/w/index.php?title=File:Hietaniemi_beach.jpg *License*: unknown *Contributors*: User:Miksa76 at en.wikipedia
File:Itäkeskus shopping centre.JPG *Source*: http://en.wikipedia.org/w/index.php?title=File:Itäkeskus_shopping_centre.JPG *License*: unknown *Contributors*: User:Kallerna
Image:Rostock Giebelhäuser Markt.jpg *Source*: http://en.wikipedia.org/w/index.php?title=File:Rostock_Giebelhäuser_Markt.jpg *License*: unknown *Contributors*: Aka, Darkone, Fransvannes, GeorgHH, Horst-schlaemma, Schiwago, Themedpark
File:Flagge der Hansestadt Rostock.svg *Source*: http://en.wikipedia.org/w/index.php?title=File:Flagge_der_Hansestadt_Rostock.svg *License*: unknown *Contributors*: Hansestadt Rostock
File:Rostock Wappen.svg *Source*: http://en.wikipedia.org/w/index.php?title=File:Rostock_Wappen.svg *License*: unknown *Contributors*: N3MO
file:Germany location map.svg *Source*: http://en.wikipedia.org/w/index.php?title=File:Germany_location_map.svg *License*: unknown *Contributors*: User:NordNordWest
File:Red_pog.svg *Source*: http://en.wikipedia.org/w/index.php?title=File:Red_pog.svg *License*: unknown *Contributors*: Anomie
Image:Rostocker Stadtbestätigung 1213.png *Source*: http://en.wikipedia.org/w/index.php?title=File:Rostocker_Stadtbestätigung_1213.png *License*: unknown *Contributors*: Kresspahl, N3MO, Schiwago
File:Plan von Rostock gez 1780-1790 J.M. von Tarnows.jpg *Source*: http://en.wikipedia.org/w/index.php?title=File:Plan_von_Rostock_gez_1780-1790_J.M._von_Tarnows.jpg *License*: unknown *Contributors*: Tarnow
Image:MAP Rostock 1910.jpg *Source*: http://en.wikipedia.org/w/index.php?title=File:MAP_Rostock_1910.jpg *License*: unknown *Contributors*: N3MO, Schiwago
Image:Bundesarchiv Bild 183-H28029, Rostock, Marktplatz mit Rathaus.jpg *Source*: http://en.wikipedia.org/w/index.php?title=File:Bundesarchiv_Bild_183-H28029,_Rostock,_Marktplatz_mit_Rathaus.jpg *License*: unknown *Contributors*: AnRo0002, Schiwago
Image:Rostock Wappen.svg *Source*: http://en.wikipedia.org/w/index.php?title=File:Rostock_Wappen.svg *License*: unknown *Contributors*: N3MO
Image:Rostock Wahlspruch.jpg *Source*: http://en.wikipedia.org/w/index.php?title=File:Rostock_Wahlspruch.jpg *License*: unknown *Contributors*: Axt, Gepardenforellenfischer, Man vyi, Schwalbe
Image:Rostock Rathaus.jpg *Source*: http://en.wikipedia.org/w/index.php?title=File:Rostock_Rathaus.jpg *License*: unknown *Contributors*: Aka, Darkone, Horst-schlaemma, Ronaldino, Schiwago
File:Flag of Norway.svg *Source*: http://en.wikipedia.org/w/index.php?title=File:Flag_of_Norway.svg *License*: unknown *Contributors*: User:Dbenbenn
File:Flag of Bulgaria.svg *Source*: http://en.wikipedia.org/w/index.php?title=File:Flag_of_Bulgaria.svg *License*: unknown *Contributors*: User:SKopp
File:Flag of France.svg *Source*: http://en.wikipedia.org/w/index.php?title=File:Flag_of_France.svg *License*: unknown *Contributors*: Anomie
File:Flag of Croatia.svg *Source*: http://en.wikipedia.org/w/index.php?title=File:Flag_of_Croatia.svg *License*: unknown *Contributors*: User:MaGa

File:Flag of the People's Republic of China.svg *Source*: http://en.wikipedia.org/w/index.php?title=File:Flag_of_the_People's_Republic_of_China.svg *License*: unknown *Contributors*: User:Denelson83, User:SKopp, User:Shizhao, User:Zscout370

File:Flag of Denmark.svg *Source*: http://en.wikipedia.org/w/index.php?title=File:Flag_of_Denmark.svg *License*: unknown *Contributors*: User:Madden

File:Regiopole Rostock Verflechtungen.jpg *Source*: http://en.wikipedia.org/w/index.php?title=File:Regiopole_Rostock_Verflechtungen.jpg *License*: unknown *Contributors*: User:Afrlmmr

File:Panorama1.jpg *Source*: http://en.wikipedia.org/w/index.php?title=File:Panorama1.jpg *License*: unknown *Contributors*: Axt, Lauchi

Image:Rostock Heiligengeisthof.jpg *Source*: http://en.wikipedia.org/w/index.php?title=File:Rostock_Heiligengeisthof.jpg *License*: unknown *Contributors*: Aka, Darkone, Schiwago

Image:Rostock Kröpeliner Straße.jpg *Source*: http://en.wikipedia.org/w/index.php?title=File:Rostock_Kröpeliner_Straße.jpg *License*: unknown *Contributors*: Aka, Darkone, Schiwago

Image:W´münde Altstadt (03) 2006-09-21.JPG *Source*: http://en.wikipedia.org/w/index.php?title=File:W´münde_Altstadt_(03)_2006-09-21.JPG *License*: unknown *Contributors*: User:Norbert Kaiser

Image:Rostock22.jpg *Source*: http://en.wikipedia.org/w/index.php?title=File:Rostock22.jpg *License*: unknown *Contributors*: Lauchi, Schiwago, Stunteltje

Image:Rostock Universität 1.jpg *Source*: http://en.wikipedia.org/w/index.php?title=File:Rostock_Universität_1.jpg *License*: unknown *Contributors*: Aka, AnRo0002, Axt, Darkone, Horst-schlaemma, Schiwago

Image:Rostock Theater im Stadthafen.jpg *Source*: http://en.wikipedia.org/w/index.php?title=File:Rostock_Theater_im_Stadthafen.jpg *License*: unknown *Contributors*: Aka, Darkone, Ronaldino, Schiwago

Image:Rostock Kempowski Archiv.jpg *Source*: http://en.wikipedia.org/w/index.php?title=File:Rostock_Kempowski_Archiv.jpg *License*: unknown *Contributors*: Aka, AndreasPraefcke, Darkone

Image:Ostseestadion.JPG *Source*: http://en.wikipedia.org/w/index.php?title=File:Ostseestadion.JPG *License*: unknown *Contributors*: Florian K, OHVChris75, Schiwago, Sebastian Wallroth

Image:Rostock-hbf-nord.jpg *Source*: http://en.wikipedia.org/w/index.php?title=File:Rostock-hbf-nord.jpg *License*: unknown *Contributors*: Chep87, Kneiphof, Platte, Schiwago, 1 anonymous edits

Image:ArosaBlu.JPG *Source*: http://en.wikipedia.org/w/index.php?title=File:ArosaBlu.JPG *License*: unknown *Contributors*: Original uploader was Jimbo Herndan at en.wikipedia

File:Liniennetz Rostock 2012.svg *Source*: http://en.wikipedia.org/w/index.php?title=File:Liniennetz_Rostock_2012.svg *License*: unknown *Contributors*: User:LuWe

Image:Sunset over Rostock harbour.jpg *Source*: http://en.wikipedia.org/w/index.php?title=File:Sunset_over_Rostock_harbour.jpg *License*: unknown *Contributors*: Maurice

Image:190407 Star Helsinki.jpg *Source*: http://en.wikipedia.org/w/index.php?title=File:190407_Star_Helsinki.jpg *License*: unknown *Contributors*: user:Kjet

File:Paldiski lighthouse 2009.jpg *Source*: http://en.wikipedia.org/w/index.php?title=File:Paldiski_lighthouse_2009.jpg *License*: unknown *Contributors*: User:Kulmalukko

File:Flag of Paldiski.svg *Source*: http://en.wikipedia.org/w/index.php?title=File:Flag_of_Paldiski.svg *License*: unknown *Contributors*: User:Joey-das-WBF

File:Paldiski_coatofarms.png *Source*: http://en.wikipedia.org/w/index.php?title=File:Paldiski_coatofarms.png *License*: unknown *Contributors*: Flying Saucer, Senzeichi, Valentinian

File:Paldiski_location.png *Source*: http://en.wikipedia.org/w/index.php?title=File:Paldiski_location.png *License*: unknown *Contributors*: User:Siim

File:Paldiski abandoned military buildings.jpg *Source*: http://en.wikipedia.org/w/index.php?title=File:Paldiski_abandoned_military_buildings.jpg *License*: unknown *Contributors*: User:Fragwürdig

File:Paldiski jaamahoone2 (2011).jpg *Source*: http://en.wikipedia.org/w/index.php?title=File:Paldiski_jaamahoone2_(2011).jpg *License*: unknown *Contributors*: User:Vamps

File:Linnavalitsus.IMGP6694.JPG *Source*: http://en.wikipedia.org/w/index.php?title=File:Linnavalitsus.IMGP6694.JPG *License*: unknown *Contributors*: User:Avjoska

File:Paldiski 2006-2.jpg *Source*: http://en.wikipedia.org/w/index.php?title=File:Paldiski_2006-2.jpg *License*: unknown *Contributors*: Bestiasonica, Dmitry G, Kulmalukko, Pomeranian, Raul6

File:Stena line logo.svg *Source*: http://en.wikipedia.org/w/index.php?title=File:Stena_line_logo.svg *License*: unknown *Contributors*: Philphos

Image:Green Arrow Up.svg *Source*: http://en.wikipedia.org/w/index.php?title=File:Green_Arrow_Up.svg *License*: unknown *Contributors*: AutisticPsycho2, Dbenbenn, Juiced lemon, Korg, Multichill, Pagrashtak, Redrose64, Trisreed, ZeroOne, 17 anonymous edits

Image:Stena Scandinavica in Kiel Harbor.jpg *Source*: http://en.wikipedia.org/w/index.php?title=File:Stena_Scandinavica_in_Kiel_Harbor.jpg *License*: unknown *Contributors*: User:Sechmet

Image:Stena Jutlandica in Gothenburg.jpg *Source*: http://en.wikipedia.org/w/index.php?title=File:Stena_Jutlandica_in_Gothenburg.jpg *License*: unknown *Contributors*: User:Marcusroos

File:Stena line route map.svg *Source*: http://en.wikipedia.org/w/index.php?title=File:Stena_line_route_map.svg *License*: unknown *Contributors*: User:Philphos, User:RedHotHeat, User:San Jose

Image:Stena Traveller.JPG *Source*: http://en.wikipedia.org/w/index.php?title=File:Stena_Traveller.JPG *License*: unknown *Contributors*: User:Chris0693

Image:Stena Traveller 2.JPG *Source*: http://en.wikipedia.org/w/index.php?title=File:Stena_Traveller_2.JPG *License*: unknown *Contributors*: User:Chris0693

File:MS Superfast VII, Neste Oilin edustalla matkalla Naantaliin Turun Korjaustelakalle, 14.5.2008.JPG *Source*: http://en.wikipedia.org/w/index.php?title=File:MS_Superfast_VII,_Neste_Oilin_edustalla_matkalla_Naantaliin_Turun_Korjaustelakalle,_14.5.2008.JPG *License*: unknown *Contributors*: User:Makele-90

MIX
Papier aus verantwortungsvollen Quellen
Paper from responsible sources
FSC® C105338

Printed by Books on Demand GmbH, Norderstedt / Germany